Lecture Notes in Networks and Systems

Volume 71

Series Editor

Janusz Kacprzyk, Systems Research Institute, Polish Academy of Sciences, Warsaw, Poland

Advisory Editors

Fernando Gomide, Department of Computer Engineering and Automation—DCA, School of Electrical and Computer Engineering—FEEC, University of Campinas—UNICAMP, São Paulo, Brazil

Okyay Kaynak, Department of Electrical and Electronic Engineering, Bogazici University, Istanbul, Turkey

Derong Liu, Department of Electrical and Computer Engineering, University of Illinois at Chicago, Chicago, USA, Institute of Automation, Chinese Academy of Sciences, Beijing, China

Witold Pedrycz, Department of Electrical and Computer Engineering, University of Alberta, Alberta, Canada, Systems Research Institute, Polish Academy of Sciences, Warsaw, Poland

Marios M. Polycarpou, Department of Electrical and Computer Engineering, KIOS Research Center for Intelligent Systems and Networks, University of Cyprus, Nicosia, Cyprus

Imre J. Rudas, Óbuda University, Budapest, Hungary

Jun Wang, Department of Computer Science, City University of Hong Kong, Kowloon, Hong Kong

The series "Lecture Notes in Networks and Systems" publishes the latest developments in Networks and Systems—quickly, informally and with high quality. Original research reported in proceedings and post-proceedings represents the core of LNNS.

Volumes published in LNNS embrace all aspects and subfields of, as well as new challenges in, Networks and Systems.

The series contains proceedings and edited volumes in systems and networks, spanning the areas of Cyber-Physical Systems, Autonomous Systems, Sensor Networks, Control Systems, Energy Systems, Automotive Systems, Biological Systems, Vehicular Networking and Connected Vehicles, Aerospace Systems, Automation, Manufacturing, Smart Grids, Nonlinear Systems, Power Systems, Robotics, Social Systems, Economic Systems and other. Of particular value to both the contributors and the readership are the short publication timeframe and the world-wide distribution and exposure which enable both a wide and rapid dissemination of research output.

The series covers the theory, applications, and perspectives on the state of the art and future developments relevant to systems and networks, decision making, control, complex processes and related areas, as embedded in the fields of interdisciplinary and applied sciences, engineering, computer science, physics, economics, social, and life sciences, as well as the paradigms and methodologies behind them.

** Indexing: The books of this series are submitted to ISI Proceedings, SCOPUS, Google Scholar and Springerlink **

More information about this series at http://www.springer.com/series/15179

Marek Szelągowski

Dynamic Business Process Management in the Knowledge Economy

Creating Value from Intellectual Capital

 Springer

Marek Szelągowski
Systems Research Institute
Polish Academy of Sciences
Warsaw, Poland

ISSN 2367-3370   ISSN 2367-3389   (electronic)
Lecture Notes in Networks and Systems
ISBN 978-3-030-17140-7   ISBN 978-3-030-17141-4   (eBook)
https://doi.org/10.1007/978-3-030-17141-4

Library of Congress Control Number: 2019936296

This Springer imprint is published by the registered company Springer Nature Switzerland AG
The registered company address is: Gewerbestrasse 11, 6330 Cham, Switzerland

Acknowledgements

The author thanks Marek Michałowski and Alejandro De La Joya Ruiz De Velasco for their support and constant focus on the practical aspect of the implementation of process management and the role of common sense. The author is grateful to Christoph Ruhsam for his great inspiration to take a broader view of process management. And last but not least, the author also thanks Witold Chmielarz, Bogdan Stefanowicz, Peter Fingar, Krzysztof Kietzman, and Jan W. Owsiński for their encouraging words, critical remarks, and patience in countless contacts and conversations during the book's creation.

Contents

About the Author

Dr. Marek Szelągowski is experienced in business process management practices. The author of the increasingly more popular concepts of "dynamic business process management" (dynamic BPM) and "Process Relevance Criterion." His innovative approach to the management of business processes is gaining more and more supporters.

He has more than 25 years of experience in implementing IT solutions in support of management. He has participated in the creation and implementation of IT solutions in the fields of accounting, human resources management, production, IT infrastructure management, etc. As the CIO of the Budimex Group in 2000–2008, he was responsible for the accommodation of digitization strategies to the changing needs of the business sector. He was managing and participating in analyses and optimizations of business processes and developing IT architecture management processes in over 75 enterprises from the construction, telecommunications, pharmaceutical, and healthcare sectors, including the Medical University of Łódź, the Ministry of Finance of the Republic of Poland, Merck Polska, EUROLOT, VECTRA, PGNiG, GK PGE, and many others.

His research interest includes: process management (including adaptive case management) and knowledge management.

He is currently employed in the Systems Research Institute of the Polish Academy of Sciences in Warsaw.

Abbreviations

ABPD	Automatic Business Process Discovery
ACM	Adaptive Case Management
ACMS	Adaptive Case Management System
AI	Artificial Intelligence
APM	Agile Project Management
APQC	American Productivity and Quality Center
BI	Business Intelligence
Big Data	Big Data
BPEL	Business Process Execution Language
BPEL4WS	Business Process Execution Language for Web Services
BPI	Business Process Improvement
BPM	Business Process Management
BPMM	Business Process Maturity Model
BPMN	Business Process Model and Notation
BPMS	Business Process Management System
BPO	Business Process Orientation; Business Process Outsourcing
BPR	Business Process Reengineering; Business Process Redesign
BSC	Balanced Scorecard
CCM	Customer Communications Management
CM	Case Management
CMM	Capability Maturity Model
CMMN	Case Management Model and Notation
CMS	Case Management System
CoP	Community of Practices
CRM	Customer Relationship Management
CSFs	Critical Success Factors
DCM	Dynamic Case Management
DMN	Decision Model and Notation
DMS	Document Management System
dynamic BPM	Dynamic Business Process Management

EBM	Evidence Based Medicine
EMR	Electronic Medical Record
EPC	Event-driven Process Chain
ERP	Enterprise Resource Planning
HIS	Hospital Information System
iBPMS	Intelligent Business Process Management Suites
IC	Intellectual Capital
ICT	Information and Communications Technology
IoT	Internet of Things
ISO	International Organization for Standardization
IT	Information Technology
KE	Knowledge Economy; Knowledge-based Economy
kiBP	Knowledge-intensive Business Process
KM	Knowledge Management
KPI	Key Performance Indicators
ML	Machine Learning
MRP	Material Requirements Planning
MRPII	Manufacturing Resource Planning
OMG	Object Management Group
PDSA	Plan–Do–Study–Act (Deming cycle)
pKM	Process-oriented Knowledge Management
RPA	Robotic Process Automation
SCM	Supply Chain Management
SECI	Socialization—Externalization—Combination—Internalization
TIQM	Total Information Quality Management
TQM	Total Quality Management
UML	Unified Modeling Language

List of Figures

List of Tables

Introduction

Modern economy is undergoing accelerating, multifaceted changes tied to the growing need of clients for more ease of access to personalized products and services. The pace, but also the qualitative character and the unpredictable nature of the undergoing changes result in fundamental principles of management with respect to the specialization of personnel and organizational units becoming insufficient [1, p. 1]. The aim of this work is to analyze the changes to process management in recent years and—on this basis—to identify a new approach to this phenomenon with a view to raising the probability of the adoption of process management in organizations and pointing to its further perspectives for development.

In the course of writing this book, the author attempted to narrow his focus to dynamic business process management and its consequences for business. For this reason, the content of this book rests on two assumptions:

1. The readers have fundamental knowledge on management, and process management in particular.

    And should they believe their knowledge to be insufficient, they remain eager to learn. For this reason, the book does not fully present the results of research into particular phenomena. It does not present the views of many distinguished authors, who either defined or analyzed specific terms and their mutual relations, etc. The book references numerous sources, which present scholarship in the discussed fields in detail, compare the methods of specific authors, and discuss their assumptions and definitions, their methods of holding research, and achieved results and conclusions, etc. The references encompass both academic sources and commercial or even advertising sources, provided that such sources present information pointing to relevant innovations, emerging directions of development, benefits for clients, or contain views or research results which are at odds with the currently effective paradigm—and for this reason unavailable in academic sources.

2. The readers have come into contact with the practice of implementing modern concepts of management, and process management in particular.

As before, in Chaps. 3 and 4 on the implementation of process management and knowledge management, the author assumes that the readers are either knowledgeable or willing to actively broaden their knowledge on the methodologies of agile process management and, in particular, on projects with a view to creating and developing IT systems and methodologies for the implementation of process management created and actively developed by consulting companies and vendors of IT solutions.

The work presents the following hypotheses:

- Modern process management is dynamic in nature.
- Traditional process management is a specific exception to dynamic process management, which is used with success in business contexts which do not require knowledge management in the course of process execution itself.
- Organizations are able to maintain a relatively lasting competitive advantage only by systemically integrating (dynamic) process management with knowledge management.

The main research questions resulting therefrom are:

1. Which goals should be set before process management in the knowledge economy?
2. Can traditional process management be used in the knowledge economy?
3. Is it possible to expand traditional process management in order to meet the requirements of the knowledge economy?
4. Is it possible to systemically manage knowledge hidden in the processes executed in the organization?
5. Is it possible to integrate (dynamic) process management and knowledge management, as well as encompass hidden knowledge within this integration?
6. Is it possible to describe dynamically managed business processes?
7. Is it possible to hold simulation and optimization research with respect to dynamically managed business processes, including unpredictable processes?
8. Should a dynamic business process have the same form of description throughout its entire life cycle?
9. Is it possible to unify process management with case management?

In order to answer the abovementioned research questions and either confirm or falsify the posed hypotheses, the author performed an elaborate, multiyear sequence of studies, which were constantly confronted with practical knowledge and in effect sometimes pointed to potential new research fields and goals. As in the case of all worthwhile theories—answers to the posed research questions lead to further questions. The answers stated in the course of the work unambiguously demonstrate that even in a turbulent business environment theoretical reflection may not only follow practice, but also show future directions of development. In effect, it may prove beneficial to observe and research the undergoing changes, even if in effect the views and paradigms which even yesterday were still perceived as dogma should simply be discarded [2, p. 18]. However, there is always the chance that new views and paradigms will arise to take their place, which will be better suited to describing reality, and perhaps will even allow us to predict the directions of further development. At the same time, they directly influence praxis, e.g., the organization of projects dealing with the implementation of process management, the use of machine learning or artificial intelligence, the method of describing knowledge-based processes, or the architectural principles of systems supporting process management.

"Dynamic business process management is not just an extension of the classical concept of process management, but also an attempt to harmonize process management with the concept of the learning organization. This is achieved through the ongoing verification of acquired knowledge with respect to the needs of the clients by numerous process performers, which leads to the gradual accumulation and proliferation of such knowledge" [3, p. 45]. In effect, knowledge is the equal-footing partner of the business process. However, even the best knowledge will not in itself lead to certain outcomes. It is action which initiates the process, or, to be more precise, individuals operating within a process "Processes don't do work, people do" [4]. In a similar fashion, it is impossible to design a good process and adapt it to the undergoing changes without access to and the proper maintenance of knowledge. In the case of numerous practical solutions, descriptions of business processes and knowledge, as well as the way they are being processed, have already converged. Business environments and solutions, in which knowledge and processes are managed separately, are fast becoming obsolete and cannot compete with organizations which empower their employees to combine the two [5].

Chapter 1 of the book contains an analysis of the evolution of process management based on books, articles, materials from the Internet and online databases, and the author's own experience from participant and non-participant observation and active, long-term involvement in implementation work. It demonstrates how the rejection of outdated paradigms opened up new directions of development. Should that not be the case, process management would still look like production management in Ford's company in the beginning of the twentieth century. It would probably have become a part of industrial automation, since in the case of line production and simple manual labor humans have long been replaced with robots. The chapter also demonstrates how under the pressures of business process management has undergone the subsequent first, second, and third stages of development, increasingly quickly accommodating itself to the multifaceted changes tied to the growing need of clients to have more ease of access to products and services. Sometimes, the direction of this development was set by theoretical works, such as those of, e.g., Porter [6] or Davenport [7], and at other times—as the result of trial and error: by way of experiments held consciously or by a stroke of luck. However, sometimes—like in recent years—business either turned a blind eye to the changes in its environment or presented as groundbreaking changes normal innovations to technologies used to date (e.g., RPA). However, the pressure of business is strong enough, and challenges set before business, including financial challenges, interesting enough, that practitioners have made and continue to make attempts at finding new solutions in a situation in which traditional process management no longer encompasses as much as 70–80% of processes in the organization operating in the knowledge economy. This requires process management to make another qualitative step in development; one which undoubtedly deserves to be named the fourth stage of development of process management. Its main assumption is the extension of process management to enable the broadest possible, daily use of intellectual capital as a source of competitive advantage. Analogous to the previous stages, this stage is first visible in practical solutions using new possibilities offered by ICT technologies, as well as in changes to social culture and work culture.

According to the accepted logic, the next step—in the second chapter—was to define the requirements and the extension of traditional process management, which would correspond to the demands of the knowledge economy. The chapter introduces the author's own concept of dynamic business process management (dynamic BPM). It has been formulated by way of posing subsequent hypotheses, adding, researching, and using Ockham's razor on subsequent conditions and principles—only seemingly necessary to allow for the use of not only the knowledge, but also the energy, the drive to action, or even the laziness of the knowledge workers executing processes in the organization. The drive behind embarking on this project was prosaic and purely practical. In a company in which the author worked as Chief Information Officer (CIO), increasingly better, more advanced IT systems supporting business were being implemented, with the caveat that the results of the implementations were becoming increasingly worse. At first, the aim of the study was simply to explain the cause behind this phenomenon. It is only after the realization that business is not static and requires support in the form of dynamic actions, which is not offered by the implemented management and IT systems, did the author begin to work on the concept of dynamic process management. Only in the course of further studies did it become obvious that one effect of process execution, which is as important as the product or service offered to the client, is value in the form or created and distributed new knowledge in the organization. An important element of the conducted research was participation in business projects, as well as analysis of unconventional industry cases, based on interviews with experts and practitioners, surveys, interviews, and in-depth interviews. A wider reflection on the technologies and IT systems used has made it possible to cooperate with various design and research teams, as well as suppliers of business process management system (BPMS) and adaptive case management system (ACMS). Despite the wide variety of cases, the analyses have nonetheless led to:

- shared conclusions with respect to the necessity and inevitability of abandoning traditional, static methods of process management in favor of dynamic methods integrated with knowledge management;

- the formulation of a definition of dynamic process management and the evaluation of the main consequences of its use, in terms of the theoretical approach both to process management in the knowledge economy and to the praxis of dynamic process management in the organization, with a particular focus on the period of implementing dynamic BPM;
- the formulation of conclusions on the practical reunification of BPMS and CMS (ACMS in particular), as well as on the theoretical compliance of both groups of systems with the formulated definition of dynamic BPM.

By applying the principle of the economy of thought, the author reduced the initially considered principles of extending traditional process management to just three, with the aim of allowing organizations to function properly in the knowledge economy, but also—for the organizations to still contain current knowledge in the scope of traditional business process management [8, pp. 50–51].

The use of these three, relatively simple, principles resulted in the consequences discussed in subchapter 2 in Chap. 2, which allowed for the generation of tangible benefits from the intellectual capital of the organization. In the author's own experience, these are changes to the business process life cycle and changes to the goal of implementing process management in the organization. The second benefit in particular stresses the direction of the development of IT systems supporting process management. This not only pertains to the use of new solutions, such as acquiring information from, e.g., the Internet of things (IoT), Big Data, process mining, robotic process automation (RPA), or machine learning, but also to an approach in which the most crucial aspect of process management is the goal of the process, because it is with this goal, and not the process diagram, that the needs and expectations of the clients (and the bottom line of the organization) are tied. Studies performed by, e.g., Gartner and Forrest unambiguously point to the fact that proponents of case management, which is often viewed as "non-process-based," were in the right. As demonstrated in Chap. 2, not only is this approach process-based, but also multiple times better in designing and executing dynamic (unpredictable, unstructured, ad hoc, etc.) processes than the traditional approach. It also allows for the much broader and the much more natural use of tools enabling machine learning or the use of elements of artificial intelligence. At the same time, in accordance with the principles of dynamic process management, it allows for the empowerment of the personnel to acquire and distribute knowledge—including tacit knowledge, which is often hard to verbalize and codify. Such knowledge is broadly revealed in practice, or during process execution, provided that the process performer is given the chance (or better still: is obligated to) to draw on such knowledge.

As will be discussed in Chap. 3, the next step was to draw on literature and case studies—from the field of health care in particular—with a view to formulating a proposal on the principles of integrating dynamic process management with knowledge management. Studies have confirmed the full compliance of the three different models of knowledge management with the model of dynamic process management. The stages of process management, such as the distribution of tacit knowledge or the creation and verification of ideas, are not separate actions within the organization, but normal components of business process execution. The proposal of updating the process-based model of knowledge management presented in the chapter underlines the possibility of it being directly tied with the model of dynamic business process management. As demonstrated, it is also harmonious with the knowledge management model by Nonaka and Takeuchi. In both instances, dynamic process management supplements the existing occasional possibilities of acquiring, uncovering, and codifying knowledge with the option of its ongoing revealing, creation, use, and verification. From the perspective of knowledge management, dynamic process management is an endless source of practical knowledge in the organization.

In order to make use of the chance offered by the expansion of process management to dynamic process management, one needs to implement it. In this regard, dynamic process management is also an extension of traditional process management. For this reason, Chap. 4 discusses the principles of and includes practical guidelines on implementing both management concepts. In the first part of the chapter, an analysis of standard methodologies for implementing traditional process management was performed, emphasizing in particular the already used elements of information or knowledge management as well as limiting the use of new information technologies. In its summary, a critical analysis was carried out showing the basic gaps in the methodologies of traditional process management (e.g., limited possibilities of using machine learning or artificial intelligence) and formulating the necessary changes resulting from it. In the next part of the fourth chapter as defined by dynamic BPM as an extension of traditional process management, no new, separate methodology has been formulated, but extension of the existing methodology to enable its use in the knowledge economy. The implementation of dynamic process management primarily requires organizations to change the goals and the scope of implementation. They must *de facto* encompass knowledge management as an essential support of the knowledge workers. For this reason, it is crucial to change the method of identifying and presenting process descriptions. Because processes may be adapted to individual contexts, to facilitate the work of process performers, it is essential to allow for the form of description of the process descriptions to change in the course of its life cycle. The studies performed by the author have demonstrated that it is possible to describe and simulate dynamic processes and that it is necessary to allow for the change of the form of description in IT systems in different stages of the process life cycle [9]. In effect, the truly "process-centered" view that hierarchy and process diagrams are the only and the best form of description is obsolete.

The work presents the author's original, theoretical concept of dynamic business process management, which expands traditional process management in accordance with the requirements of the knowledge economy. It has been developed by the author in the course of his collaboration with vendors of systems supporting process management (BPMS and ACMS), as well as teams executing commercial implementation projects. However, despite efforts to maintain theoretical cohesion, the work was primarily written with practicing managers in mind, who in the undergoing changes see opportunities of which they wish to take advantage. However, they do not desire to "reinvent the wheel"—due to both the risk of failure and the time required. For this reason, they attentively observe the realities of conducting business and seek out possibilities of the most efficient use of the dynamism and the knowledge of their coworkers. They may feel that fashionable technologies, such as robotic process automation (RPA) or process mining, have limited process optimization capabilities and cannot provide (relatively) sustainable competitive advantage. As long as artificial intelligence is not really intelligent, a sustainable competitive advantage can only be achieved by using people's intelligence and engagement, much more difficult to perform and less spectacular than automating routine operations by purchasing applications for RPA or process mining.

References

1. Płoszajski P (2004) Organizacja przyszłości: przerażony kameleon. W kierunku nowej filozofii zarządzania. Retrieved from http://www.allternet.most.org.pl/SOD/Heterarchia%20prof._Ploszajski_-_Organizacja_przyszlosci.pdf. Accessed 22 April 2017

2. Rifkin J (2016) Społeczeństwo zerowych kosztów krańcowych. Internet przedmiotów, Ekonomia współdzielenia, Zmierzch kapitalizmu (The Zero Marginal Cost Society: The Internet of Things, the Collaborative Commons, and the Eclipse of Capitalism). Wydawnictwo Studio EMKA, Warszawa

3. Czekaj J (2009) Metody zarządzania procesami w świetle studiów i badań empirycznych. Uniwersytet Ekonomiczny w Krakowie, Kraków

4. Brown J, Gray E (2005) The People Are The Company. Retrieved from http://www.fastcompany.com/magazine/01/people.html. Accessed 10 Dec 2017

5. Taylor C (2012) Reunifying Knowledge and Business Process Management. Retrieved from http://citeseerx.ist.psu.edu/viewdoc/download?doi=10.1.1.225.9570&rep=rep1&type=pdf. Accessed 18 July 2017

6. Porter M (1985) Competitive advantage. Free Press, New York

7. Davenport T, Prusak L (1998) Working knowledge—how organisations manage. What they know. Harvard Business School Press, Boston

8. Heller M (2014) Granice nauki. Copernicus Center Press, Kraków

9. Gottanka R, Meyer N (2012) ModelAsYouGo: (Re-)Design of S-BPM process models during execution time. In: Stary C (ed) S-BPM ONE 2012. LNBIP, vol. 104, Springer, Heidelberg, pp 91–105

Introduction to the Second Edition

The book *Dynamic Business Process Management in the Knowledge Economy: Creating Value from Intellectual Capital* was primarily written with managers in mind, who in the undergoing changes see opportunities of which they wish to take advantage. For this reason, this is a book on innovations emerging to a larger degree from practice rather than theoretical, academic reflection. It challenges the paradigm of traditional process management, which stems from as far back as industrial engineering [1], by describing modern process management, which is entirely different from that from 50, 10, or even 5 years ago. Perhaps this sounds suspiciously "non-technical," but by using the new possibilities offered by developing technologies and IT tools, it is focused on creating conditions for the development of the individual and teams of people through work. This, of course, is not due to the fact that owners and managers in organizations have suddenly become philanthropists, but because during work it is easiest to make natural use of existing tacit knowledge, as well as to create and verify in practice new knowledge arising from employees, teams, or the entire organization. Among the numerous names for methodologies, concepts, and proposals for the direction of developing process management in literature, such as *agile*, *intelligent*, *adaptive*, or *human*, the name **dynamic** has been chosen to underline that the actual source of all further possibilities offered by dynamic business process management is the dynamism of the knowledge workers. Not just their knowledge, but also their will to take action determines whether the knowledge workers will introduce agile, intelligent adaptations in the course of performance itself, with a view to raising process efficiency, often in the form of minor adaptations to the specific context of performance. Such adaptations derive from tacit knowledge, as well as the engagement and the will of the knowledge workers, thanks to which new knowledge is created and verified (according to the proverb "necessity is the mother of invention") in the course of process performance itself. The concept of dynamic business process management does not replace the concept of traditional business process management, as much as it expands it, allowing for the creation of an organization which does not limit or hinder employees in their work, but rather, on which provides the necessary conditions for the broadest possible use of their knowledge and dynamism.

The book discusses differences between the methods and tools of implementing static (structured) and dynamic (semi-structured and unstructured) processes, as well as their consequences. It explains why the extremely disruptive developing technologies, such as robotic process automation (RPA), have limited potential in regard to raising process efficiency by replacing routine human work with faster and error-free work performed by software robots, albeit work limited to repeatable, routine static processes. It explains why, until artificial intelligence (AI) does not in fact become actually intelligent, constant competitive advantage in organizations will only be achievable thanks to the use of the knowledge, intelligence, and engagement of people. This is much harder than implementing, e.g., a "paper-free office," the purchase of an RPA, implementing cloud servers, or undertaking other actions which are beneficial thanks to the simple use of ICT. The dynamic business process management solutions presented in the book are less spectacular, but can be used in the case of any processes, and imbue organizations with a foundation for competitive advantage which is hard to copy—one based on the use of the entire knowledge and dynamism of the employees thanks to the emerging integration of process management and knowledge management, which already at present enables the increasing use of technologies and IT systems allowing for process mining or machine learning (ML).

Reference

1. Taylor FW (1911) The principles of scientific management. Harper & Brothers, New York

Chapter 1
Traditional Business Process Management

Abstract The rapid development of process management and its practical uses stems from the changing conditions of business, which are the result of overlapping and mutually stimulating changes in business culture, social conventions, the development of information and communication technologies, as well as the process of globalization and changes to the principles of competition themselves. For several years now it has become apparent that practical methodologies and IT systems supporting the implementation and use of process management in organizations are developing at a much faster pace than their theoretical underpinnings. This chapter discusses the limitations of traditional business process management, as well as its main discrepancies with the requirements of business in the knowledge economy. The 3rd wave of development of process management, which has been initiated around the year 2003, is becoming increasingly less responsive to the requirements of modern business. There is a lack of theoretical reflection on traditional process management, despite the fact that due to changes in the paradigms of the knowledge economy it may be used in the case of a mere 20–30% of the processes within the organization. The chapter describes changes to process management and the practical solutions which exceed traditional process management. Of fundamental significance for practitioners dealing with the preparation of tools and the implementation of methodologies pertaining to process management is the distinct integration of the methodologies and tools of business process management (BPM) and case management (CM). Such integrations are often the result of trial and error, sometimes fall outside of the mainstream of process management, but have nonetheless been positively verified by the clients, that is, in the most objective manner possible. The chapter also points to the factors leading to the emergence, as well as the main characteristics of the so-called 4th wave of process management. Just as in the case of the previous waves, its direction is mostly determined by practical solutions.

Keywords Business process management (BPM) · Dynamic business process management (dynamic BPM) · Case management (CM) · Knowledge management (KM) · The 3rd wave of BPM · The 4th wave of BPM

© Springer Nature Switzerland AG 2019   1
M. Szelągowski, *Dynamic Business Process Management in the Knowledge Economy*, Lecture Notes in Networks and Systems 71,
https://doi.org/10.1007/978-3-030-17141-4_1

1.1 Introduction

Modern economy is undergoing accelerating, unforeseeable changes, which are the result of overlapping and mutually stimulating changes in the business culture, social conventions, the development of information and communication technologies, as well as the process of globalization and changes to the principles of competition themselves. Both the rapid pace of development, as well as the qualitative character of the undergoing changes, result in the necessity of searching for new solutions, which would fundamentally change current fields and principles of business management. The characteristics of this new order and their business implications still remain unknown. Undoubtedly, process management remains a crucial part thereof, as it adapts the management of organizations to the conditions set by both the global market and individual consumers at a faster pace than other concepts. However, is traditional business process management, which distances the decision on the method of performing work from the act of performance itself, capable of being used in the knowledge economy? Is the feedback loop of adapting processes to the changing conditions of performance too long and too slow in the hypercompetitive environment? In an economy in which the only constant is change, can organizations which depend on the knowledge and engagement of a mere fraction of their personnel, which makes decisions on the shape of the performed processes, maintain constant competitive advantage? The aim of this chapter is to discuss the limitations of traditional process management, which make it unresponsive to the requirements of business. The chapter also discusses attempts at adjusting process management to the needs of the knowledge economy, as well as the possibilities (and challenges) offered by using new ICT solutions and their methodologies, such as case management, machine learning, process mining, robotic process automation, *Big Data* analysis, or the Internet of Things (IoT).

1.2 The Development of the Concept of Business Process Management

Modern economy is undergoing accelerating, multifaceted changes, which are tied to the growing needs of the customers with respect to easier access to individualized products and services. Both the pace, as well as the qualitative character and the unpredictable nature of the undergoing changes result in standard principles of management, which are rooted in the specialization of workers and departments, becoming obsolete. This, in turn, necessitates the search for new solutions, which would fundamentally change current fields and principles of business. The characteristics of this new order still remain unknown. Furthermore, the characteristics themselves change due to the unending:

- changes in business culture and social conventions;
- globalization and changes in the principles of competition;
- development of information and communication technologies.

The above list of factors necessitating changes to the methods of management in organizations is not itself exhaustive. Furthermore, all such factors are mutually stimulating in a manner which is hard, if not impossible, to foresee. However, it is generally assumed that organizations which meet the new expectations will be "slimmer," flatter in terms of their structure, more focused on the work of teams, flexible, continually adapting their actions to the needs and requirements of their clients [1].

It is becoming increasingly more apparent that solutions which were once (perhaps even the day before) something to be proud of should now be seen in terms of a failure in regard to the needs of the client, who is searching for products with diverse features, available on demand, within a reasonable price range, and of perfect quality. Organizations must be committed to working toward meeting both requirements voiced in the past, as well as emerging, often vaguely foreseen requirements of the future. They must remain on top of the changing needs of their clients. To this end, companies should analyze the habits and choices of their clients on an ongoing basis. However, it is no longer possible to understand the clients' ongoing needs on the basis of evaluating their past choices. Changes in client habits resulting from globalization, technological changes, and the implementation of scientific innovation (in e.g. IT and medicine) are so rapid that it is becoming crucial to gain an understanding and work in the present itself, on the basis of knowledge on the foreseeable future [2, p. 1]. This requires organizations to make a daily effort to adapt their principles of operation and update their knowledge on the present and future requirements of their clients. It has become necessary to constantly broaden knowledge, as well as collect and analyze experiences resulting from ongoing contacts with clients, partners, and even competitors. In short, organizations must constantly learn how to take action. They must stay on top of rapid changes in terms of both technology and their surroundings, and, first and foremost, stay on top of the growing individual needs of their clients. Companies are forced to raise the appeal of their offer not just by lowering costs and improving efficiency, but also through the personalization of products and services. The aim is to approach the new ideal: the single-client market. On this market, there is no optimal or ideal method of managing business [3, p. 48].

Competition is not only becoming more fierce, but also more rapid [4]. Besides traditional competitive fields, such as design and quality, competition is fast becoming a test of speed. According to the spokesperson for Boeing, Scott Griffin, "a fast company beats a slow company every time" [3]. The founder of the Yankee Group, Howard Anderson, believes that "companies must now have «one overriding goal: speed. Speed at all costs... hyper-speed.»" [5]. To paraphrase, a company which will manage to adapt itself and its products to the requirements of its clients at a faster pace than the competition will gain competitive advantage regardless of its current state of resources and possibilities. Fixed assets and expansive organizational structures are no longer a benefit, but merely an operational cost; one which might lead to failure. In large organizations with extensive multilevel structures speed is forfeit due to the slow pace of the decision-making process, the need to hold internal negotiations, the need to accept changes to the budget, etc. [6, pp. 45–57]. Preparing the organization to meet the conditions of the modern knowledge economy requires us to abandon the "foresee and control" approach in favor of "act and learn." The aim of a management strategy is to define the direction (a strategic vision), as well as to prepare the organization and lead it into the uncertain and ambiguous future, so that it maintains its competitive advantage in terms of pace, speed, and initiative. This model is based not on a single correct answer, but on the concept of the constant, gradual active learning by observing the environment, including its reactions to the tactical operations of the organization. While forecasts are a good decision-making tools, strategic scenarios are better as tools for shaping tactics. Instead of a definite "yes/no" scenario, they provide the organization with a means of building and developing tactics as an iterative process (by means of multiple repetitions).

Recent years have witnessed multiple "expensive surprises," which had an enormous influence on entire sectors of the economy. They have demonstrated that organizations without access to cutting-edge management methods were unable to use the opportunities in their surroundings to their advantage. They were unprepared for the emergence of sudden qualitative changes, which were not included in broadly accepted predictions. They had no knowledge management system which would enable them to collect, interpret, and react to delicate anticipatory signals coming from their environment [7]. Instead of—like Whirlpool—creating a system of collecting and developing ideas emerging from existing organizational structured, multiple companies closed their innovation process in ghettos, different incubators, funds for special projects, business development departments, and teams dedicated to "outflanking" the competition, detaching innovation from the fundamental operations of the enterprise [8]. It is, therefore, not surprising that their reaction to changes was usually belated, and oftentimes simply mistaken. With the growing pace of hypercompetition and growing dynamic strategic interactions between competitors on the global market, companies require the implementation of new management methods and tools [9]. Methods which allow for flexible adaptation and innovation supported by the ongoing measurement of the needs of the clients, the capabilities of the competition, and the efficiency of the company's processes, as well as tools which allow the company to influence and stay on top of the changing needs of the clients by way of broad, unrestrained experimenting and the rapid implementation of the acquired knowledge in the fundamental operation of the enterprise. This requires the top management to understand that the management of the entire knowledge within the organization and the fast pace of organizational learning are the foundation of competitive advantage [10].

Since the beginning of the 20th century, we can distinguish three, or perhaps four clearly different stages of the development of process management, which in relevant literature are often called "waves of process management" [11, pp. 18–23]:

I. Industrial engineering   (1911)
II. Value chain management   (1985)
III. Evolutionary adaptation to the needs of the clients   (2003)

The above stages illustrate the methods of adapting process management to the requirements set by changes to business, as well as changes in the approach to management itself.

1.2.1 The First Wave of Process Management – "Industrial Engineering"

The main goal of process management in the 1st wave was better time utilization, cost reduction, and expanding the production volume. This approach was used in the analysis and improvement of the functioning of production processes. The main assumptions of this approach were:

- an understanding of the process as a sequence of actions describing subsequent work operations;
- the division of the process into fundamentals, or even the atomization of work [12];
- the elimination of redundant actions and unnecessary losses;
- the complete expendability of workers performing simple, simple tasks;
- the elimination of initiative, both innovative and performative, among workers.

The Ford construction line has become the symbol of industrial engineering. It necessitated workers executing as "industrial robots" to execute a specific sequence of action within a specific time range in order to meet the tempo of production. Toffler [5, pp. 290–293] writes: "Work was «de-skilled» or dumbed-down, standardized, broken into the simplest operations." Despite the fact that Taylor [13] was the first to connect knowledge on working with working itself, he was also the one who separated thinking from action, setting in stone for a long time the division between managers ("thinkers") and those executing the actual work ("doers"). In this concept, "[a] worker is a kind of organic robot, operated by a manager via remote control" [14]. In the first half of the 20th century, the use of industrial engineering has allowed companies for the increase in the efficiency of manual workers by an order of magnitude. The most benefits went to those companies which were the first to introduce the concept (e.g. Ford Motor Company) or were able to coordinate all of their operations to perfection (e.g. Toyota).

However, due to the separation of thinking and performance, the concept of industrial engineering has been burdened with two crucial faults:

- It allowed for raising the efficiency of manual labor, but it failed in regard to mental work. The latter requires creative thinking due to its very nature and it is impossible to compartmentalize it into simple performing actions which would require no thinking whatsoever. For this reason, it is also impossible to automate, which in the scope of industrial engineering would mean full repetition regardless of the individual personnel performing the work required.

- It requires both the technological process and the product itself to remain unchanged for long durations. If management, planning, quality control, etc., are separated from the actual performance of work, the proper preparation and coordination of the production process requires a long time due to the very precise nature of the analysis of all of the possible scenarios in order to defend the organization from all probable or simply possible threats. This, in turn, means that it is impossible to introduce rapid changes to the product line, nor rapid changes of a single product to the requirements of the client.

Industrial engineering yielded great results at a time when:

- the main problem was the organization of manual labor;
- technologies were practically unchanging and work was performed in stable conditions;
- the product was standardized and shipped without taking into account the individual needs of the client.

However, following the growing needs of the clients and the growing pace of changes introduced to both products and services the principles of industrial engineering had ceased to provide a competitive advantage. They had started to become a burden, or even the cause f failure, at a time of increased market change and increased customer interest in access to a diverse range of products. A good example of the erosion of the benefits provided at first by the implementation of the principles of the 1st, stage of process management is the history of the Ford Motor Company [15]. The company has been producing a single model of an automobile for 20 years, which in the years 1914–1926 has also been offered in a single color. However, what provided considerable competitive advantage in the years 1908–1920 almost lead to the company's downfall in the years 1920–1927 [16]. The response to changing requirements in the practical dimension of business management was the 2nd stage of process management.

1.2.2 The Second Wave of Process Management – "Value Chain Management"

The main goal of the 2nd wave of process management took the form of management focused on the value offered to the client. Its main assumptions were:

- The total operations of an organization should be focused on providing products and services to the client. All of the processes within the organization should be subordinated to this task;
- Each action or group of actions should provide value to the client;
- The value is dependent not only on the quality of work performed in the course of specific actions or their groups, but on their coordination as well.

Porter's research on the value chain [17] is considered to be the beginning of the 2nd wave of process management. In 1986, Deming [18] formulated the *Deming Flow Diagram*, which described the horizontal connections within a vertical organization, running from the supplier to the client, as a process which can be measured and improved upon. In their article *New industrial engineering: Information technology and business process redesign*, Davenport and Short [19] stated that the principles of process orientation are fundamental to the organization. In the same year, Hammer [20] presented the concept of process orientation as the fundamental component of reengineering. Such a proliferation of meaningful and often game-changing works in the span of a single decade demonstrates the rapid pace of changes and the strong pressure of business to adjust management to the changing rules of competition and conditions of operation.

The main directions of the development of process management in the course of the 2nd wave of process management can be divided into revolutionary and evolutionary concepts and methodologies.

Revolutionary Concepts and Methodologies

The best-known among the concepts of revolutionary changes to managing the organization is the concept of Business Process Reengineering [21]. It calls for:

- the radical redesign of the organization and its processes;
- the rejection of existing principles of action (in regard to the organization and its processes) with the aim of a sharp rise in efficiency, followed by a rise in profit, by 50, 100%, or more;

However, the actual execution of those principles has led to:

- significant problems with the management of human resources, stemming from the irrational expectation of a abrupt changes in organizational culture;
- the loss of a significant part of the knowledge of the organization.

Most attempts at implementing Business Process Reengineering—exceeding 70%—have ended in failure. At present, no more concepts and methodologies are introduced with the mindset of a "fundamental rethinking and a radical redesigning of the organization" and there seem to be no practical implementations of the concept of Business Process Reengineering [22, 23].

Evolutionary Concepts and Methodologies

Evolutionary concepts and methodologies were based on various modifications and practical extensions of the Deming PDCA (or rather, PDSA) Cycle. They called for:

- the ongoing, evolutionary innovation within the organization;
- the operation of the organization in a way which enables constant innovation.

The most representative among the evolutionary concepts is the concept of Business Process Redesign [19, 24]. However, despite multiple successes, with the rising pace and the unforeseeable character of changes in the market economy, implementations stemming from this group of concepts have begun to fall short of the requirements that organizations started facing in effect of their:

- over-focus on perfecting instead of innovating;
- lack of openness to radical changes caused by e.g. new, groundbreaking technologies or rapid changes to social life and work culture;
- slow and limited approach to knowledge management.

Process management concepts within the 2nd wave of process management have developed under the influence of the needs and expectations of business, and, first and foremost, the experiences gained during subsequent implementation. The cause of this was the growing pressure of doing business, caused by:

- the growing volatility and pace of operations;
- globalization;
- the rapid development of information and communication technology;
- changes to social culture as the result of e.g. widespread digitization of life and work, which forced organizations to search for new methods of operation and, as a result, new methods of management.

This constant practical verification of a theoretical process management concept pointed to the cumulating changes, which have become the basis of formulating the theoretical framework for the 3rd wave of process management.

1.2.3   The Third Wave of Process Management – "Evolutionary Adaptation to the Needs of the Clients"

The main goal of the 3rd wave of process management is to enable organizations to adapt to the changing needs of their clients on an ongoing, evolutionary basis. Its main assumptions are:

- process management as a coherent and flexible system of operation and innovation within the organization;
- management of the entire process from the point of view of the client—with the engagement of the organization's suppliers and partners;
- the harmonious use of ICT technologies in order to raise the quality of management (flexibility speed, accessibility mobility, transparency, etc.) and shorten the process optimization loop [11].

The 3rd wave of process management stresses the importance of using information and communication technologies in order to ensure day-to-day business flexibility, all the while keeping in mind the necessity to delegate powers in places which are close to the place of operation. Unfortunately, the assumptions of the 3rd wave did not include the needs and the attempts of using process management to manage unstructured processes, which have been going on since the nineties, also with the use of methodologies and tools of case management [11]. As a result, the proposed solutions still do not encompass the accelerating individualization of the needs of the clients, nor the growing significance of creativity and inventiveness on the part of knowledge workers. With changes to the social culture and work culture due to widespread digitization of work and life alike, such factors have a growing influence on the management of modern organizations. At the same time, the paradigm of superiority of a process diagram over different forms of process description, which is promoted by a large number of scholars and practitioners of process management, results in most organizations becoming unable to implement process management in accordance with 3rd wave principles [25, p. 10].[1]

The proposed evolution of process management in the course of the last 100 years demonstrates how fundamental changes in requirements and expectations of holding business resulted in changes to conceptions and practical rules (sometimes even implicit unwritten rules) of implementing process management. Table 1.1 includes the main factors forcing the initiation of the subsequent waves in the development of business process management. The factors which in accordance with predictions will not be satisfied under the 3rd wave of process management have been preceded with a question mark.

The direction of the development of process management is determined by two fundamental, mutually stimulating and strengthening factors:

- the development of information and communication technologies;
- changes in work and life social culture.

Both resulted in the rapid acceleration and globalization of business operations as the result of the rapid acceleration and expansion of the possibility to share knowledge and the de facto dismantling of the barriers that distance once put up for holding business. Due to their rapid, widespread adoption in the private sphere, information and communication technologies are also used without cultural barriers in business. New business technologies and models, such as the personalization of products and services, e-commerce, mobile technologies, social media applications, cloud computing, the Internet of Things, and elements of Artificial Intelligence have practically entered business at the same time as entering the private lives of millions, if not billions of potential workers and clients. In effect, it is impossible to ignore new factors which necessitate introducing changes to process management in organizations, such as:

- the growing digitization of business;
- the growing individualization of processes with the use of *Big Data* techniques, robotic and automation, and artificial intelligence;
- further changes to social norms due to widespread digitization of work and life (forced digitization).

---

[1]"Flowchart-based technologies treat work as steps in a routinized process, and hence people as cogs in a machine." [25, p. 10].

Table 1.1 Factors behind changes and changes in approach to process management

Wave of process management		Rules (assumptions)	Fundamental change factors
I	Industrial engineering (1911–1980)	• No process changes or slow pace of process changes • Elimination of redundant actions and unnecessary losses • Division of the process into simple elements • Full expendability of workers performing simple tasks	• Larger product and service variability, which necessitates larger production process variability • Growing significance of intellectual work • Growing focus on services
II	Value chain management (1985–2003)	• Each task or group of task must provide value for the client • The value is dependent not only on the quality of the work performed in the course of different actions or their groups, but also on their coordination as well • Processes within the organization are innovated upon through evolutionary or revolutionary means	• Globalization • Growing volatility and pace of operations • Rapid development of common ICT technologies
III	Evolutionary adaptation to the needs of the clients (2003–2017)	• Process management as a cohesive and flexible system of operation and innovation within the organization • The entire process is being managed from the point of view of the client, while also taking into consideration the organization's suppliers and partners • Harmonious use of information and communication technologies (ICT) in order to raise the efficiency of management and shorten the process optimization loop	• Changes to social culture due to the common digitization of work and life (forced digitization) • Growing digitization of business • Required individualization of processes with the use of *Big Data* techniques and Artificial Intelligence • Growing importance of knowledge and the practical use of intellectual capital for the organization

Source Author's own elaboration

1.3 The Incompatibility of Traditional Process Management with the Requirements of Business

Traditional business process management assumes that process performers are not authorized to introduce changes to the process in the course of performance itself [26]. In effect, these processes are static from the perspective of the process performers. Their course should run in accordance with predefined algorithms ("musical scores," as musicians would say) in the course of a description or model. The performance of traditional processes should resemble the performance of a symphonic orchestra ensemble. In order to achieve the goal, all musicians must perform in accordance with the score prepared beforehand by the composer and in the same tempo indicated by the conductor. Each deviation from the musical score blemishes the final effect—the musical performance. There is no freedom to introduce changes to the music in the course of performance itself, nor to improvise on the part of the musicians or the conductor.

Descriptions (models) of traditional business processes change at a much slower pace than the time required for their performance, which allows us to conclude that the process does not undergo changes in the course of performance itself. It is possible to improve processes on the basis of standard mechanisms, which usually stem from the concept of Edward Deming and are based on different modifications of the PDSA cycle [18, 24, pp. 34–40, 27]. Optimization, periodic process improvement, or building a process-driven organization do not change this implicit assumption. The above method of implementing process management has been immensely successful in times when changes to the market, the principles of competition, or client requirements have been introduced over much longer periods of time than the length of the process and did not require an individual approach to clients. With the rising pace of changes, it became impossible to perform processes with an average, standard client in mind. In effect, implementing and using traditional process management in the organization has been faced with increasing problems.

Traditional, static implementations of process management provide, among others, the following benefits:

- revealing and evaluating knowledge in the organization, usually in the form of a one-off initiative in the Process Identification stage;
- the transparency and availability of published and binding process maps and models;
- reducing costs of process execution thanks to automation;
- full control over process execution both in their course and upon their completion (instant capture of mistakes and deviations from process performance).

At the same time, however, the implementation of traditional process management is tied to certain risks, among which the most severe are as follows:

- the threat of becoming detached from the changing market conditions;
- generating losses as the result of performing the standard variant of the process, which does not conform to the specific conditions of performance;

- no option of process individualization;
- creating or strengthening a corporate culture of no accountability.

Below are selected problems facing implementations of traditional business process management.

1.3.1 Lack of an Actual Process-Based Approach to Operations

The vertical structure of modern organizations has been created in order to increase control and make better use of work specialization. In practice, however, it often leads to abandoning the fulfillment of client needs as the main objective of the organization. Boundaries between organizational units (the so-called silos or dens) result in the compartmentalization of the fundamental process, which generates value for the clients, into disparate, disharmonious parts [5]. Optimization, or rather, the suboptimization of processes within the confines of such units may result in individual organizational units attaining their goals independently of the organization's fundamental goal, in a way which does not add value, or which is even at odds with the bottom line of the organization as a whole. In the case of traditional business process management this threat is severe, because the standard course of a given process is usually authorized by individuals who are managers of specific organizational units and who are responsible for their results. In such circumstance, it is challenging to make a strong effort to introduce changes to processes which would improve the results of the organization as a whole, and which would not improve the results of a specific organizational unit or which might even result in their becoming worse. It is even harder to accept the initiatives of one's subordinates with a view to improving the operations of the organization as a whole. Instead, managers will be much more willing to accept initiatives with the aim of improving the operations of a specific ("our") organizational unit. The lack of a genuine process-based approach to implementing traditional business process management may not only lead to diminishing results in the organization as a whole, but also result in lowering the personnel's pace of adaptation and a drop in their motivation.

The chief financial officer (and a member of the management board) of one construction company was so fixated on reaching his department's own goals, which consisted of generating the most return on investment from deposits and other financial instruments, that the financial department stopped paying the subcontractors and suppliers on term, which in turn led to enormous losses due to halted construction on one of the essential contracts due to the mass resignation of subcontractors from cooperation!

1.3.2 Detaching the Decision on the Method of Performance from Performance Itself[2]

Business process descriptions define the methods of process performance within the organization. Detailed descriptions (in the form of rules and regulations or procedures), tables, diagrams, or process models define both the course (the logic) of the process, as well as the responsibilities of its particular participants, the documents and data used, IT systems facilitating process performance, as well as the goals and key performance indicators, etc. They are prepared by a process-modeling team prior to process performance, that is, without specific knowledge on the client, the specific context of performance, and often without regard for the experience of the individuals assigned to a given process. Decisions on the final form of process descriptions are made by process owners, who are usually not responsible for the day-to-day performance of the processes in question. In effect, they do not possess full knowledge on the actual, genuine course of the process, but instead act upon their own perception thereof, which is based on their own experiences from years past, on information devoid of context, or—in the worst case scenario—on a vision of how the process should look. This leads to the process performers being forced to adopt a specific form of description, which is incompatible with the actual course of the process prior to performance itself (and sometimes as early as when it is being authorized by the management). In result:

- processes are executed in accordance with imposed rules and regulations, resulting in less value for the organization—or even in net loss;
- the best employees who are penalized for deviating from procedure and performing processes in a nonstandard manner can lose their motivation or quit altogether;
- employees perform processes by adapting them to the actual conditions of performance, but hide this fact from their supervisors. This leads to the emergence of a knowledge gap between process performers and the management (the so-called "hidden factory" effect) [28]. Without actively counteracting this phenomenon, this gap might become even broader. Since the management is not aware of

---

[2]For the purposes of this work the author has adopted a division into creative and reproductive executions, which correspond to the English terms:

- *execution*—the reproductive, passive execution of a chain of commands prepared and transmitted prior to execution (analogous to e.g. the preparation of a computer program in the form of a chain of commands in an EXE file (EXE from *EXEcution*). Even when the employee (*executor, doer*) sees mistakes in the program, he or she cannot change the program in the course of execution. This would require making changes to the source code itself, another compilation, and executing the new version of the program.)
- *performance*—the creative, active execution of work with a view to achieving a specific goal. Of course, such work should still make use of prepared standards, diagrams, or guidelines, but its performance is determined by the performer (analogous to performing a jazz improvisation in accordance with musical canons and thanks to previous preparation, albeit in a creative manner resulting from i.a. the context of performance, e.g. the reaction of the public).

the actual course of processes, it believes itself to be infallible and continues to "optimize" the processes in a direction which results in diminishing efficiency.

> In a medium-sized company from the metal sector near Warsaw, in the course of a consulting project with the aim of initiating the implementation of process management with process identification, during a meeting with the management—after having modeled logistical and production processes with a team of employees—the CEO of the company insisted the processes be remodeled or supplemented with additional initial events. Such an events were described as the "decision of the CEO" and were to be given the highest priority in terms of triggering the "process" in accordance with his decisions. The only argument in favor of this inclusion had been: "I'm in control and as we can see—we are doing just fine." The implementation has been discontinued.

1.3.3   The Overlong Process Improvement Loop

In the case of traditional business process management, the first stage of implementing process management consists of process identification [29, p. 22, 31]. All models of process maturity as the first step toward process management define "process identification and standardization" [30, 31]. They usually take the form of a one-off project, after which in accordance with some variation of the Deming cycle (PDSA) experiences are collected at shorter or longer intervals, after which improvements to process businesses are implemented. This mechanism is present both in evolutionary methodologies (e.g. Business Process Redesign), as well as in methodologies proposing revolutionary changes to business processes with a view to accommodating the organization to the new market environment (e.g. Business Process Reengineering). In both cases, we are dealing with improvement feedback loops overseen by the management, with the difference pertaining to the strength and the radicalism of the first stage of introducing changes. In both cases, decisions are made by an implementation committee or the management of the organization not in reaction to the actual ongoing needs arising in the course of process performance, but as the result of planned cycles of audits (e.g. ISO) or process overviews, which are instituted annually or—in the best case scenario—quarterly. A situation in which the pace of changes is comparable with (or lower than) the pace of the improvement feedback loop puts the organization at risk. In such circumstance, it is impossible to stay on top of change. The newly implemented solutions will always be overdue in light of the situation at hand and the needs of the process performers. Of course, it is downright impossible to use the improvement mechanism when the client needs are individualized, nor in the case of technologies such as machine learning.

In 2005, one of the largest construction companies in Poland had implemented procedures pertaining to the management of procuring construction materials and services, which necessitated obtaining at least three offers from potential suppliers with a price fix of at least 30 days, etc. During the construction materials crisis, when prices fluctuated daily by a dozen, and weekly even several dozen percent, the procurement department personnel have maintained the constant procurement of goods and services by contracting with the suppliers that the conditions of the contracts, including the price and the amount, be agreed upon by phone on a case-by-case basis for each specific construction project and for specific delivery and payment dates. The conditions of each offer were immediately confirmed via e-mail and the order was placed on the same day. The personnel of the procurement department did not even attempt to petition the management in order to change the official procedure, as that would take at least a month (or two weeks in the minimum). The change would take too much time—time which was needed to maintain the constant procurement of construction materials.

1.3.4   The Incapability of Process Individualization

Modern businesses either already mass produce goods and services tailored to their customers' particular needs (mass customizing) or will have to learn how to mass produce such goods and services in the near future [1, p. 8]. At the same time, business processes in these companies pertaining to e.g. construction projects, on-line sales, or health diagnostics, have to be carried out with different means in accordance with the customers' (investors', purchasers', recipients', patients', etc.) specific requirements. Because it is impossible to predict and model all potential courses of patient treatment, types of projects, or all of the different customers' potential requirements, the need arises for the dynamic adaptation of business processes to the individual conditions of each particular performance. In such circumstance, the implementation of traditional, static process management in an organization cannot be limited to the routine, fully repeatable performance of the same, even the most expertly optimized standard process. In turn, the implementation of traditional, static process management means that following the process identification an improvement stage, the organization works in accordance with such standard, averaged processes. Any deviations therefrom require the approval of owners, process leaders, the management, or the directorship, etc. Due to the necessity of maintaining a pace of operation required by the clients and the number of concurrent process execution, it is impossible for the process owner to analyze, decide on, and introduce changes to processes on an ongoing basis (e.g. large construction companies in Poland perform over a hundred concurrent investments, consulting companies hold several hundred projects and provide several hundred opinions annually, and legal firms take on hundreds of different cases, not to mention medical doctors performing multiple diagnostic-therapeutic processes at the same time in the case of different patients with the same condition, but in different clinical states).

At the same time, it is very probable that the requirements to individualized processes set by different groups of clients or individual clients themselves may be mutually exclusive. Analysis of the required changes may very often lead to the conclusion that it is impossible to design a process which would be compliant with the expectations of all of the clients (there are no e.g. ideal diagnostic-therapeutic processes). Then, the only option left in the case of traditional process management is to:

- design disparate versions of the process for different groups of clients or individual clients. (Due to time constraints and matters of procedure it is impossible to constantly design and approve new versions of a given process for each client. Apart from rising costs, this may also lead to informational chaos.)
- average the requirements and design the process with an "averaged" client in mind. (This in turn might require the organization to drop multiple clients with individual needs or—in the case of diagnostic-therapeutic processes—be downright impossible.)

In the aviation industry it is impossible to design a standard operational process for crisis situations. Flight controllers and airship captains operate under standard ASSIST (Acknowledge, Separate, Silence, Inform, Support, Time) or ANC (Aviate, Navigate, Communicate) guidelines and standard checklists suggesting which actions should be undertaken in situations similar to those described in the guidelines (e.g. the European 4444 directive). However, it is the controllers' sole discretion to decide which actions will be undertaken and in what sequence [32].

1.3.5 Diffusion of Accountability for the Results of Process Execution

From the perspective of a specific performer, a process description contains information on the method of performance which has been approved and imposed by the organization's management. In this regard, the process performer is to a considerable degree limited in the freedom to decide on the course of the process and is forced to proceed in accordance with a predefined algorithm, even when in the knowledge that in situation at hand such actions are sub-optimal, lead to losses, the loss of a client, strengthening the position of the competition, etc. The decisions on the method of process execution made by the management de facto prevent process performers from using their knowledge to adapt the process to the conditions of a specific performance with a view to its optimization, or at least avoiding losses. But because it is not the process performers who are authorized to make decisions on the course of the process, they are consequently rid of the accountability for the results of the performed processes. It stands to reason that accountability should rest in those individuals who have imposed the method of performing work. At the same time, the process owners who were responsible for deciding on the method of performing processes do not feel accountable for each single performance thereof. In effect, a situation may arise in which upon completing the process identification stage, no one feels responsible for poor results or losses resulting from a specific performance. This is a source of severe problems with implementation, and the implementation of traditional business process management in particular.

1.3.6 A Top-Down Approach to Implementing Changes

One of the main accusations directed at both evolutionary process-based method-ologies (e.g. Business Process Redesigns), as well as methodologies proposing rev-olutionary changes to business processes (e.g. Business Process Reengineering) is their failure to recognize the significance of necessary changes to the organizational culture and the intellectual capital of the human resources in the organization. Even if some innovations are suggested at the initiative of the employees themselves, such proposals are nonetheless verified and either accepted, rejected, or even ignored out-right by the management [33], when "[t]he process, antibureaucratic to its roots, can function only in atmosphere that gives individuals considerable autonomy" [5]. Despite declaring the significance of intellectual capital, methodologies from the field of traditional business process management clearly point to the management of the organization as the initiators, architects, and arbiters who are to decide about the implemented solutions down to the last detail [21, 22]. The largest controversies were tied to dividing accountability for process identification (between external consul-tants and the management) and implementation (the employees)—or even attempts at leaving the implementation in the hands of external consultants. In the course of implementation, such a top-down, authoritarian style of management, which objec-tifies the employees of the organization, results in:

- the lack of the employees' identification with the implemented solutions, which sometimes manifests itself in active resistance thereto;
- the lack of the employees' engagement and feeling of responsibility both during the implementation, as well as the day-to-day use of the implemented solutions in the field of process management;
- the rising risk of the projects of implementing process management becoming mismatched with the aims or the actual capabilities of the organization due to wishful thinking—or often the short-term goals—of the organization.

This often results in the detachment of the implemented solutions from day-to-day work. In consequence, the employees see them as another external reporting and controlling system, which they must cheat. In such case, a common practice is to perform the process in a manner expected by the client, but without revealing this fact to the management. This leads to the emergence of the so-called "hidden factory" effect, due to which the organization irretrievably looses knowledge on the actual causes behind both successes and failures. This effect consists in the personnel developing tacit processes and systems of operation, which in the best case scenario fix the mistakes and correct the discrepancies off official standard processes and procedures. Such actions are concealed from the management and block the possibility of introducing improvements, as in effect the management has a false awareness of the course, costs, and efficiency of the processes at hand [28].

1.3.7 No Possibility of Widespread Limited Experimentation

Traditional business process management does not offer the option of limited experiments being performed by a broad range of performers in the course of performance itself. This means that employees are unable to generate practical knowledge in the form of experience gained by process performers. At the same time, because large groups of process employees are prevented from experimenting in search of new solutions, the organization as a whole is systemically prevented from broadly using its intellectual capital. In practice, this means that the organization:

- must forego the possibility of arriving at one or more solutions as the result of creative limited experiments, which after analysis could have been adapted in multiple specific scenarios or among which the most effective solutions could have been chosen;
- must forego the possibility of holding forestalling experiments, which could have allowed it to prepare itself for future changes;
- must forego the possibility of evaluating the results of the introduced changes in order to eliminate i.e. compensatory feedback loops, or situations in which well-intentioned changes result in a response from the system which nullifies the expected benefits. This may lead to a situation in which the management is unintentionally working against the organization and itself is the reason behind new difficulties.

One of the attempts at solving this problem has the form of the limited delegation of privileges to introduce changes to process performers. However, a more thorough analysis of the effects of Japanese experiences (lean management, *kaizen*) shows that this solution generates two issues:

- the time-limited delegation of severely limited privileges means that more significant changes are still introduced at a slow pace, as they still must be consulted with and approved by the management. In turn, small changes, which have virtually no effect on a given process, will be introduced by the performers themselves at their sole discretion. Japanese experiences show that in such circumstance, there are in fact no incentives for the employees to propose significant changes or perform experiments with a view to introducing significant improvements to the process at hand. Each year, millions of proposals are filed, albeit to negligible effect [34]. This points to the risk of underestimating or rejecting, or perhaps proceeding too long over proposed significant changes on the part of the management, and the lack of such proposals resulting therefrom;

- the delegation of broad privileges to process performers without any ongoing over-sight results in organizational chaos. This usually leads to the reestablishment of "silos of power and information" and the rapid return of the problems associated with traditional business hierarchy, e.g. the collision of competences, no intercon-nection between elementary processes, the reestablishment of barriers of transit between organizational units (this time on the level of processes), the emergence of "no man's land," etc.

1.3.8 The Inability to Keep Up with the Needs of the Clients and the Changing Environment

When an organization is divided into different organizational units, it finds itself unable to identify the knowledge it owns or obtains from clients or from the mar-ket in general. Unfortunately, this is caused by company management wanting to control knowledge as strictly as i.e. finances. In effect, instead of flowing directly (and thus quickly) between process performers, knowledge meanders through the hierarchic waters of the entire organization, slowing down, becoming distorted, or even becoming lost in the process [35]. In effect, the organization's reaction time to environmental changes increases considerably. It also often leads companies to repeat the same mistakes, which are usually the result of the views of the management or continued (and costly) attempts at reinventing already-existing solutions time and again, which did not meet with the approval of the decision-makers in the past. At the same time, compartmentalization is the cause of missing out on essential oppor-tunities of gaining advantage over the competition only because such chances were not included in procedures and formal strategies and the process executors were not independent enough to try and take advantage of the opportunities presented to them.

1.3.9 The Use of a Small Fraction of the Knowledge in the Organization

In the case of traditional, static approaches to implementing process management, the aim of implementation often rests in the "optimization" or "raising efficiency" of the performed processes. It usually has the form of a one-off action, following which the intensity of process improvement quickly drops to its usual level. The above approach to implementation results in process optimization limited to removing the errors and reducing losses discovered in the modeling stage, or, in other words, limited to using knowledge available during implementation itself. Because the implementation is held and overseen by the organization's management, the members of which has the function of process "leaders" or "owners," the knowledge used corresponds to the views of the management. And since (as will be discussed in more detail in Chap. 3) such knowledge is often but a mere fraction of the entire knowledge of the organization [36, pp. 63–64],[3] the value of unupdated knowledge contained in process models fall the quicker, the quicker the changes in the organization and its surroundings. In a hypercompetitive environment, this value may even be negative, as old, outdated knowledge may result in the sub-optimal performance of processes or the performance of processes which in light of new knowledge are completely unnecessary.

> In one of the largest construction companies in Poland it took the manage-ment four attempts to acknowledge the fact that once again it had chosen a subcontractor and a waterproofing system which do not work well in prac-tice. The savings made in the subcontractor offer selection stage were several times lower than the losses incurred by the required warranty repairs (it was necessary to e.g. uncover the foundation of an inhabited building, remove the old protections, complete the repairs, install the new protections, restore the surroundings of the building to their old shape, and sometimes pay an addi-tional penalty). However, the tender procedure instituted by the management did not account for the evaluation, nor the exchange of information between different departments of the organization on the efficiency and the durability of the solution at hand. It only authorized the contract manager to seek out the lowest possible price.

1.3.10 Insufficient Support for, or Even Conflict with the Concept of Knowledge Management

Market changes and changes to individual client needs cannot be controlled or fully predicted. The organization, however, may ignore them or adapt to them. For obvious reasons, ignoring such changes and expecting that the market and the clients will themselves adapt to the requirements of the organization is not a good solution. For

---

[3]"This proves that the knowledge about business processes is neither in management nor with the business or BPM consultants but with the people who execute." [36, 29, p. 64].

that reason, organizations must work out systemic adaptive mechanisms designed on the basis of their knowledge, which is updated on an ongoing basis. The assumption held in traditional business process management that process performers have no right to introduce changes to processes in the course of performance itself is the cause of the three main problems pertaining to knowledge management in the organization:

- the lack of a systemic mechanism of creating or collecting knowledge;
- the lack of support for creative, limited experiments which result in the creation of new knowledge;
- the lack of an institutionalized expansion of the group of individuals creating innovations, which leaves such actions in the hands of the management alone.

Knowledge is not being systemically created or obtained in the course of process execution, as processes are performed in accordance with predefined descriptions. The only available option is the ex-post evaluation of the used knowledge and an attempt to change process descriptions with the use of the results of the evaluation or with the use of knowledge obtained from outside of the organization. In both cases, we are dealing with actions instituted in the aftermath of the process, which constitute additional work without effect on the results of the completed process itself. In both cases, we are also dealing with the lack of extending knowledge management with a view to encompassing the process performers. The organization is limited to using the knowledge of its management and—in the case of knowledge obtained from outside—the knowledge of external consultants. The employees themselves do not function as knowledge workers ("*performers*"), but as classical doers ("*executors*").

The above list does not exhaust all of the problems resulting from the implementation of traditional business process management. It could also include the following:

- problems with interpretation resulting from disseminating information (i.e. in the Process Identification and Improvement stages) outside of the specific context and time of process execution, which create the risk of a misunderstanding and, in effect, the risk of overestimating or underestimating the signaled problems [14, p. 38];
- problems with uncovering tacit knowledge in the course of additional actions pertaining to knowledge codification or verification instead of the course of day-to-day operations;
- limited possibilities of the systemic learning-by-doing, since actions must be performed in accordance with a predefined pattern, deviations from which are prohibited;
- the reluctance to reveal tacit knowledge obtained in the course of performance due to the threat of revealing actions which run against procedures contained in process models;
- the inability to use process mining techniques due to not including data or entering false data in event logs of the IT systems supporting process execution;
- the systemic lack of a clear goal with respect to the use of tools from the fields of machine learning or artificial intelligence in the course of process execution limited to the repetition of a predefined pattern,

as well as numerous other practical problems and limitations. Beside quantifiable losses they result in the lack of motivation and a further drop in the creativity of process performers ("even if we tell them what to do, they will not change the procedure in time"). In such circumstance, as early as in 2000 Rummler proposed shortening the improvement feedback loop in order to enable employees to improve their work on the basis of objective measurements, virtually without having to wait for the decision of the management. He assumed that "[w]e have found that about 80% of the opportunities are in in the Skills and Knowledge area. We have found that fewer than 1% of performance problems results from Individual Capacity deficiencies. Our experience is consistent with that of Deming [37], who maintains that only 15% of performance problems are worker problems and 85% are management problems." [38, p. 106]. In effect, it is essential to change the management of the workstation. Shortening the feedback loop is another term for delegating part of the privileges, which to date have been in the hands of the management, to the people who perform the actual work at hand. In turn, this requires going outside the confines of traditional business process management.

1.4   The Search for the Possibility of Adapting Traditional Business Process Management to the Requirements of the Knowledge Economy with the Support of IT Solutions

The abovementioned issues that implementations of traditional business processes management are faced with are the result of the discrepancies of traditional BPM with the requirements of the knowledge economy. They arise during attempts at implementing traditional, static process management in business, which is fast becoming increasingly more dynamic in nature. As numerous research by various organizations and research institutes demonstrates, static processes account for about 20–40% of all the processes within the organization and this percentage is constantly becoming lower [39, pp. 20–33, 41–43, p. 30, 44–46]. These processes are primarily:

- processes regulated by external laws (e.g. accounting or financial processes);
- internal processes of the organization, which have no substantive contact with client-facing processes;
- production processes highly regulated due to objective external criteria (e.g. biological, physical, chemical) or held patents, concessions, and licenses.

In the remaining 60–80% of cases, particularly those which pertain to the creation or supply of products and services dedicated to external clients, it is impossible to simply repeat in detail a once-defined "ideal" process, as there are simply no ideal processes in the knowledge economy to begin with. Each performance of a given diagnostic-therapeutic or sales process requires the awareness of the specific contest of the performance, which often cannot be foreseen [47]. As Fig. 1.1 illustrates, by

Fig. 1.1 The degree of
business processes
structurization. *Source*
Author's own elaboration, on
the basis of Di Ciccio et al.
[48, p. 2], Kemsley [49]

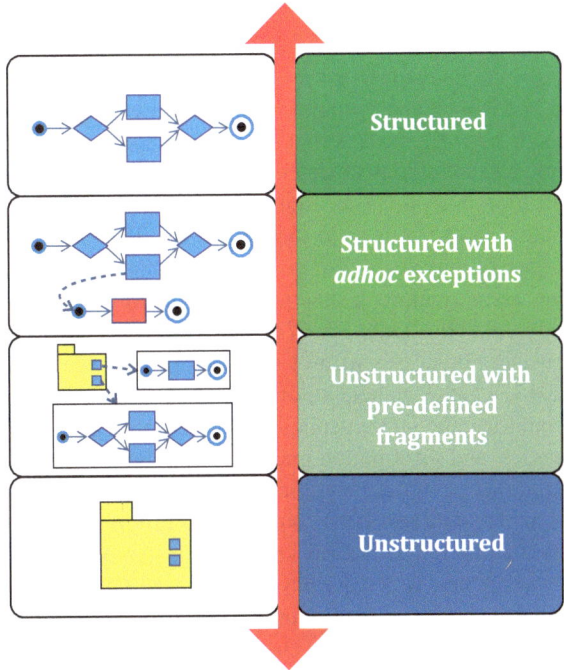

dividing processes on the basis of the degree of their structure, as recently as in 2012 it was widely believed that dynamic (unstructured, ad hoc) processes cannot be modeled at all [48].

In consequence, within traditional process management such processes are de facto impossible to manage or improve on, as according to the definition of the Deming Circle or Thomas Davenport's concept of Business Process Redesign, modeling is the first fundamental stage of management and innovation [15, 18, 24, 50]. In this case, it seems obvious that in reality most of the processes within an organization in the knowledge economy fall outside the scope of traditional business process management. This group includes processes which are critical in the knowledge economy, such as research and development processes or processes pertaining to supplying products and providing services to clients.

1.4.1 Pressure Resulting from the Development of ICT Technologies and Changes to Social Culture

Within the concept of Business Process Reengineering, Hammer and Champy [21] have underlined the close connection between the rapid development of IT and radical changes in management, though they were not the first to do so—the same connection had been made by i.e. Leavitt an Whisler as early as in 1958 [51]. However, reengineering was the first methodology in which such significance has been assigned to information technologies: at the same time treated as a challenge, a catalyst, and an essential factor of implementing radical changes in the field of management. Since then, it is becoming increasingly more apparent that the use of ICT technologies is both the solution to changes in management, as well as one of their main causes. The proliferation of ICT technologies enables the constant growth of the pace and global nature of operations. At the same time, their use is essential to the management of organizations in the increasingly hypercompetitive environment. This fact is being increasingly more acknowledged by practitioners as well. As early as in 2004, over 30% of Polish and 17% of foreign managers have agreed that they aim to combine business processes with IT solutions [52]. In the "The CEO Challenge" study from 2015, which was held among over 600 CEOs of different companies from multiple countries, the implementation of new technologies from the fields of products, processes, and information management, was the fourth task with the highest priority in the scope of implementing innovative strategies in organizations. Of significance is the fact that for Asian countries this task places third, whereas for European countries—fifth on the priority scale [53, pp. 31–35].[4] This is a self-accelerating process, in which new ICT solutions are the answer to the growing requirements of management. They open management to new possibilities, which support newer, emerging technologies—and so on. One crucial element which further strengthens the positive feedback loop are rapid changes to social culture and work culture in the direction of the day-to-day, broad use of solutions offered by ICT technologies. Just five years ago most managers and employees did not use mobile cloud computing applications on a day-to-day basis. Today, thanks to mobile devices such as tablets and smartphones, such solutions are the standard in personal life and are rapidly being adapted informally in professional life. Organizations cannot simply ignore such challenges, as the same solutions are being used by their competition as a source of competitive advantage—they may not only be crucial in terms of further development, but also survival on the market itself.

The pressure of changes within the organization and in its immediate environment is so large, and the potential for competitive advantage so high, that even the rapid pace of changes is too slow for many organizations. This especially concerns innovative organizations operating in the knowledge economy, in which most of the fundamental processes fall outside the scope of traditional, static business process management. In order to overcome this limitation, recent years saw the emergence of numerous concepts, methodologies, and technologies aimed at facilitating the management of processes which are dynamic in nature, or as dependent on such a large number of different factors that in practice it is impossible to account for all of them at once [54]. The most widely known of them are (in an alphabetical order):

---

[4]No. 1 Create culture of innovation by promoting and rewarding entrepreneurship and risk taking.
   No. 3 Raise employee engagement to drive productivity.
   No. 4 Redesign business processes.
   No. 5 Continual improvement (lean six sigma etc.) [53, p. 35].

- Adaptive Case Management;
- Adaptive Processes;
- Advance Case Management;
- Agent BPM;
- Agile BPM;
- Business Process Investment;
- Business Process Orientation;
- Business Process Reengineering;
- Business Process Renewal;
- Cognitive BPM;
- Dynamic BPM;
- Dynamic Case Management;
- Human BPM;
- Intelligent BPM;
- Process Mining;
- Social BPM;
- Subject-Oriented Business Process Management.

Of course, the list is far from complete. It should definitely be supplemented with qualitative methodologies such as TQM (and TIQM), SixSigma, or lean management, which, while admittedly not new, are now in their renaissance thanks to the development of ICT [55]. Due to the intensity of using ICT solutions and their support for knowledge management, we should point to the group of methodologies and technologies which are based on case management, process mining, robotic process automation (RPA), as well as methodologies and technologies pertaining to machine learning, deep learning and artificial intelligence.

1.4.2   Case Management—An Attempt at Surpassing the Limitations of Traditional Process Management

Case management was introduced in response to the needs of business, which were unable to be met with the use of traditional business process management. In the case of work performed in a repetitive manner, it is possible to describe (model) the course of the repeated process. Then, it is possible to optimize the process with the use of traditional process methodologies and optimize it in accordance with the Deming cycle. However, how do we support work which is not repetitive and routine in nature, and the completion of which is dependent on the decisions made mid-performance in accordance with the situation at hand, as well as the dynamism and the knowledge, including the experience, of the decision-makers themselves? How are we to describe e.g. the process of intervening in a crisis situation? Or a life-threatening scenario for the patient? Or in the case of a fire?

In all of the aforementioned situations, it is impossible to describe the procedure in the form of a detailed algorithm. The reaction is dependent on such a large number of unique circumstances that it is impossible to determine the proper course of action beforehand. However, in such circumstances it is possible to describe the resources at hand and the standard fundamental processes describing their use (a defibrillator must be charged and checked before use; the fire hose should be extended and connected to the water source before the nozzle is used, etc.). The MD overseeing a life-saving procedure will work with the use of the equipment and materials (e.g. blood, drugs, surgical dressing) at hand, and, first and foremost, in reliance on the knowledge of the participants. Likewise, the on-site chief fire officer overseeing the fire zone is reliant on the equipment, materials (e.g. water, fire-fighting foam), the context of the procedure (whether or not access to the site is obstructed, the location of water sources), as well as the skills of the on-site team. In the course of describing the performance of unforeseeable processes, we are able to describe data encompassing e.g. the resources and the context of performance, as well as those parts of the process which are foreseeable and repeatable [43, pp. 7–20, 56]. However, as the creators of case management stated as far back as in the nineties, it is impossible to prepare process descriptions in the form of diagrams depicting the sequence of actions.

With time, due to the rejection of subsequent proposals of changes to the traditional process-based approach, a separate methodology was formed; and at the same time, IT tools supporting its use were developed, with the aim of empowering knowledge workers to perform processes in accordance with available data and information, the specific context of each performance, and their own experience (e.g. FLOWer, ECHO, Staffware Case Handler) [56]. This allowed organizations to build a much more versatile work environment; one focused on innovation in comparison with that allowed by traditional business process management. The development of case management systems allowed for the introduction of transparent mechanisms of performing and monitoring work, without limitations in the form of process diagrams as imposed algorithms of performance. After the development paths of BPM and CM systems separated, two approaches to supporting the performance of business processes surfaced [57–60]:

- In business process management (BPM), the process itself is the most important [61]. The process includes data, the scope of which was described in the modeling stage, and which are used and expanded upon in the course of process execution [31].
- In case management (CM), the most important factor are structured data, or, in other words, information. It is information which describes the resources at hand and the context of work performance, as well as forms the basis of making decisions and performing actions on the part of knowledge workers, who perform work in accordance with their knowledge [56, 62].

Both approaches to routine (repeatable) work and knowledge-based work can be depicted in the form of a simple comparison (Fig. 1.2).[5]

---

[5]In the newest versions of BPMS systems, which allow for the performance of tasks on the basis of case management methodologies, the comparison depicted in Fig. 1.2 was replaced by separate perspectives on data and process description. In effect, it makes no sense to stress the difference between centering processes and centering data.

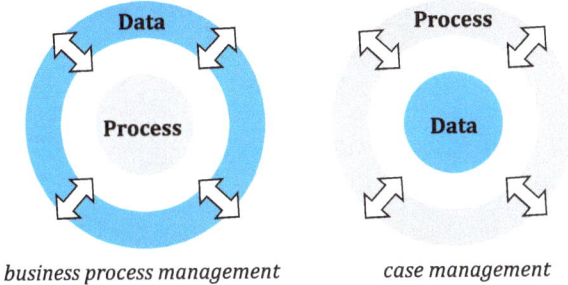

Fig. 1.2 Approach to data and process workflows in process management and case management methodologies. *Source* Swenson [63]

In both cases, we are dealing with process performance, the use of data, roles performing tasks, etc. However, traditional process management focuses on the earlier uncovering and standardization of knowledge within the organization in the form of process models. In turn, case management postulates that process performance is dependent on the specific context of performance and the knowledge of the process performance, known in this methodology as knowledge workers. Their goal is to reach business goals. In the case management approach, the process is considered completed when it fulfills its goal—unlike in traditional business process management, when it is considered finished upon completion of a specific sequence of actions prepared in the form of a process diagram (Fig. 1.3).

The performance of a case (or, to be more specific, the process of case performance) is held in the direction set in accordance with the knowledge of the knowledge worker, the results of previous performances, and the changing context of performance. It is impossible to define the course of the process upfront as a sequence of specific actions, the completion of which ends the performance of the case (or, to be more specific, the process of case performance) itself.

The goal of case management is to manage the broadest possible scope of data which influence process performance. For this reason, a crucial element of CMS systems is the management of business information, including those sent via email and those created in social media [36, pp. 20–21] (Fig. 1.4).

At present, there is no single methodology of case management. The two most popular are *adaptive case management* (ACM) and *dynamic case management* (DCM). In the case of both, case management identifies just 3 stages of business process performance:

- **Modeling**. The fundamental aim of this stage is to prepare the work environment in IT systems by cooperating with IT to define such advanced elements of the process as principles of compliance, event triggers, content elements, record formats, and ontologies defining the types of objects and their mutual relations. The modeling stage may also encompass the preparation of process patterns on the basis of the knowledge of a selected group of employees (analogous to process identification in BPM methodologies). However, much more often process patterns are

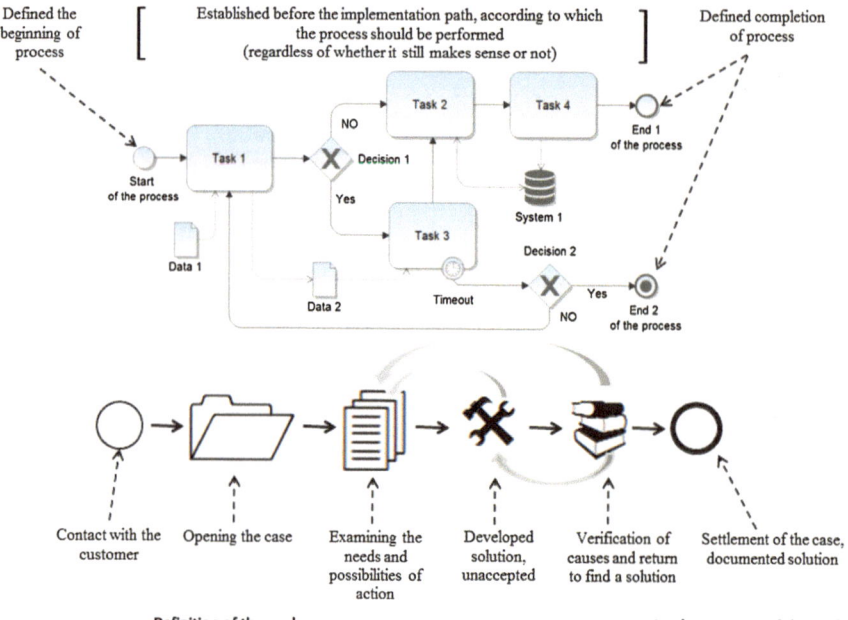

Definition of the goal   **Implementation of the goal**

Fig. 1.3 Focus on process performance and goal performance in the methodologies of traditional process management and case management. *Source* Author's own elaboration

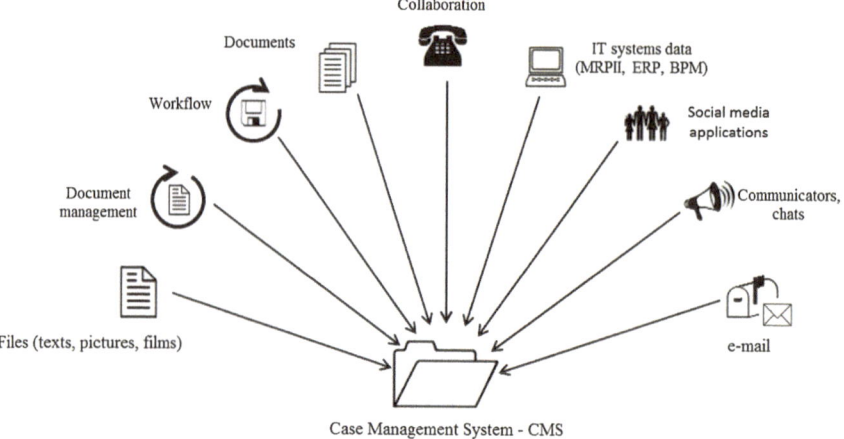

Fig. 1.4 Diverse sources of data integrated within CMS systems. *Source* Author's own elaboration

prepared in the course of evaluating process performances (cases) and establishing patterns by means of adjusting or streamlining the process to the foreseen context of performance.

- **Discovery**. The fundamental stage of business process performance, with the option of either using predefined process patters and action types or creating ones from scratch. The work is performed with the support of the IT system, but does not require the assistance of IT specialists. On this stage, data from different channels of communication (email, fax, SMS, communicators, social media applications, etc.) is automatically entered into the process. Furthermore, on this stage CMS systems record all of the actions of the process performer (the knowledge worker), as well as a broad spectrum of data describing the specific context of the performed process. (If the graphical process diagram contains 20% of the total information about the process, thanks to this mechanism case management allows for the collection of a considerable part of the remaining 80% of the information describing the context of a specific process performance.). This stage also provides the users with support in the scope of (re)Designing the decision-making process through recommendations, which might be accepted or rejected.
- **Adaptation**. This stage contains advanced options dealing with monitoring and evaluating the efficiency of the process by monitoring the methods of process performance, including engaging the knowledge workers or the clients in the process of evaluating the results of the process by e.g. voting or submitting detailed evaluations [29, pp. 31–32]. On the basis of information and employee and client evaluations obtained and submitted in the course of process performance itself it is possible to streamline process patterns and user interfaces in view of future use.

ICT systems supporting the implementation and the practical use of case management methodologies allow for the completion of work by "nontechnical" business users by creating and consolidating unstructured processes from fundamental patterns, contents, information-exchange systems, social interactions, and business principles defined in the Modeling stage, without the need of creating process diagrams. During process performance, knowledge workers using case management methodologies and tools are able to make any decisions and initiate any forms of cooperation in accordance with their level of authorization. They are not limited by decision points ("decision gates") or principles of communication ("communication flows") included in predefined process diagrams. This frees the knowledge workers to actually perform on the basis of their experiences and within the specific context of the performance [47]. At the same time, they contain ICT solutions which allow for the collection of knowledge during the entire process lifecycle—from process creation or discovery, through the process performance stage, up to the adaptation stage—as well as obtaining knowledge from completed processes. Systems supporting process performance collect contextual knowledge—skipping the intermediate analysis stage—directly from the actions of business users. ACM and DCM methodologies are based on the fact that the performance of all processes is fully transparent and controllable both during performance itself and upon its completion. Thanks to the inclusion in CMS of transparent mechanisms allowing for the ongoing monitoring and collection of information on the performed work and the broad context of its performance there is virtually no need to resort to separate techniques of process mining. The same mechanisms of monitoring and collecting knowledge allow for the real-time support of the decision-making process for both managers and all knowledge workers within the organization.

The implementation and use of case management methodologies is not always met with the approval of managers and knowledge workers. The main reservations brought up in this regard are [33, 44, 64, 65]:

- the lack of a cohesive, comprehensive case management methodology, which would describe the integration of case management with other elements of management within the organization;
- the risk of suboptimizing process performance; of streamlining processes to the requirements and goals of a specific knowledge worker instead of the organization itself;
- the threat of decision and organizational chaos resulting from the fact that repeatable processes within the organization have a unique course with each subsequent performance;
- the lack of or a severely limited scope of process identification before performance. This puts the responsibility for process discovery on the knowledge workers themselves, which is not always accepted;
- a higher risk of operation, particularly in the first stage of implementing case management, before the creation and practical verification of a process pattern library, as well as the implementation of an organizational mechanism of the creation and adaptation of process patterns to the requirements of a specific performance;
- very high requirements in regard to the skills and engagement of the knowledge workers and managers;
- the risk of using two BPMS and CMS process systems within the organization;
- the lack of a single notation system for processes within case management. (The CMMN notation published by OMG [66] in 2014 is not sufficiently popular nor mature.)

Undoubtedly, in practice the largest uncertainty is tied to the severely limited scope of process identification prior to performance. A situation is theoretically possible, in which there are no process descriptions prior to performance and the performers are forced to rely solely on their knowledge and intuition.

CMS systems also warrant further development. The main drawbacks as described by M. Pucher [43] in 2010 are:

- too much focus of the system creators on functional requirements;
- the lack of a flexible framework for configuring processes to account for external changes and internal business practices;
- the ongoing assumption that all process (subprocess) pathways can be identified in the design stage;
- the lack of flexible management in the face of the appearance of unforeseen changes to processes (subprocesses) which have been identified as being repeatable (static) in nature;

- the lack of integration between the performed process and the available knowledge;
- the lack of integration between modules, data, equipment and services.

Nevertheless, despite the above-mentioned reservations, the rapid development of CMS systems focused on the use of new, emerging technological possibilities (communicators, social media applications, elements of artificial intelligence, and *Big Data* technologies) results in case management becoming perhaps the most promising direction of the practical management of processes in the knowledge economy.

1.4.3 Process Mining

ICT tools and methodologies, which are currently known under the name of process mining, are a significant extension of IT tools which are have been in existence for over a decade, and which are dedicated to automated business process discovery (ABPD) [67, p. 51, 68, 69]. As early as in 2010 a Gartner analyst, Jim Sinur, has published study results, according to which the concept of automatic business process discovery was one of the ten most important technologies of business process improvement (BPI) [70].

The Process Mining Manifesto, which has been published in 2012, defined the goal, the scope, the requirements, and the foundations of the practical execution of methodologies and techniques pertaining to process mining. Process mining is comprised of the following methods [71, pp. 1–2]:

- automatic discovery of the workflow structure (the creation of new process models on the basis of data collected from the event log);
- conformance checking (monitoring deviations of performance from the predetermined model);
- analysis of social networks within the organization;
- automatic creation of simulation models;
- predictive analysis and real-time generation of recommendations.

The basic three types of process mining are: discovery, conformance checking, and enhancement (Fig. 1.5).

The project of implementing and using process mining in the organization consists of five main stages (Fig. 1.6):

- stage 0: Plan and justify
- stage 1: Extract
- stage 2: Create control-flow model and connect event log
- stage 3: Create integrated process model
- stage 4: Operational support

The actual implementation project begins following the business process planning and justification stage (stage 1). In stage 1, Extract, source of business data are

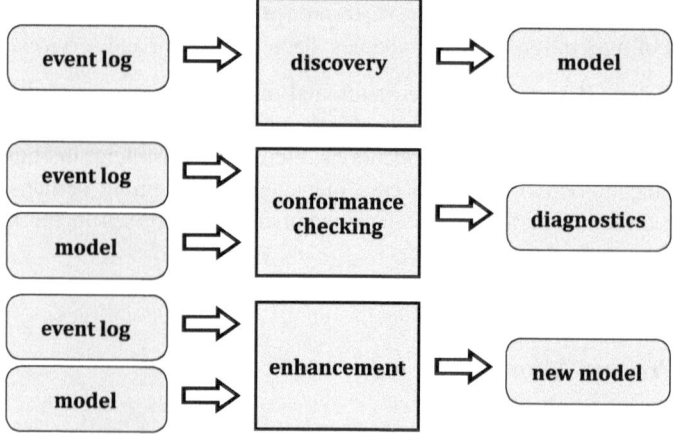

Fig. 1.5 Three basic types of process mining. *Source* Process Mining Manifesto [71, p. 4]

identified and the significance of particular groups of data and their relations is established. Furthermore, in this stage queries for the retrieval of data from selected data sources are designed. In stage 2, models of process control-flow are created with the use of selected mining algorithms. When the process control-flow model is organized and purged of events resulting from e.g. information noise present in event logs, it can be enhanced in stage 4 with additional perspectives, such as time, resources, events, etc. Such prepared models can then be used in operations, as well as enhanced or used to check the conformance of the performed processes (stage 4).

In accordance with ABPD or process mining methodologies, organizations are capable of discovering and using in practice their knowledge on the performed business processes, even when such knowledge is contained not in data from process (BPMS or CMS) or transactional (MRP II, ERP, CRM, or HIS/EMR) systems, but in e-mail networks, messaging applications, social media portals, etc. [34, p. 38]. An integral part of this concept is the possibility of the systemic inclusion of process mining in business process management as on of the variants of process identification or process "discovery," as well as process "enhancement" [67, 71, p. 4, 72]. This allows for the identification of a process on the basis of the performed work (like in the case of case management), while simultaneously enabling the comparative analysis of the performed processes with the predefined standard process—which in turn leads to the enhancement of the standard process (like in process management). In effect, it is possible to extend process analysis with data collected in traditional transactional systems, such as ERP, CRM, MRP, or communication systems (e-mail, SMS, Skype, etc.), or even social media (Facebook, Twitter, LinkedIn, Yammer, etc.), or various combinations thereof, with the use of both mobile and stationary devices [34, p. 38].

Initially, process mining focused on the analysis of historical data. However, currently we may distinguish between three directions of process mining: identification, prediction, and recommendation. One of the goals of process mining is the creation

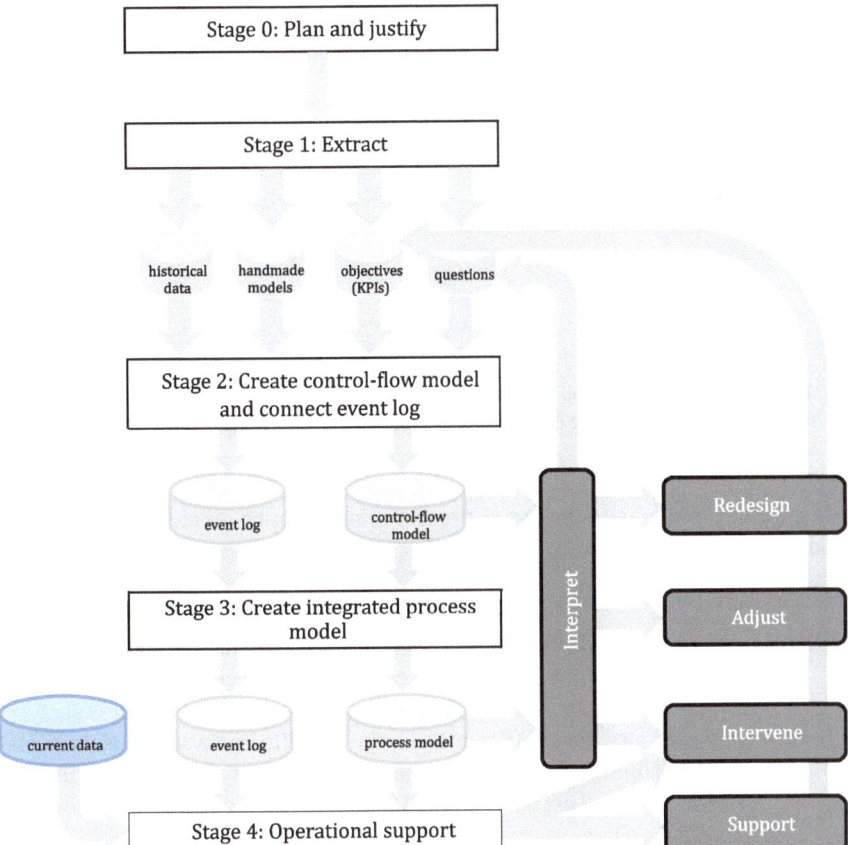

Fig. 1.6 Model of a project pertaining to the implementation and use of process mining in the organization. *Source* Process Mining Manifesto [71, p. 6]

of the so-called dynamic process model, that is, a model which may be used and updated on an uninterrupted basis, as contrasted with a static model, which is prepared in the course of a one-off analysis, after which it is shelved. New data on events may be used to uncover knowledge on possible courses of process performance (cases), which have not been accounted for in the existing model. For this reason, user interaction with the model should be accounted for in day-to-day performance.

Broad implementation and use of process mining tools and methodologies still requires us to overcome multiple challenges and practical hurdles. Figure 1.7 depicts the main challenges set before process mining tools and methodologies identified in the course of a study by Claes and Poels [73]. The color blue indicates the part of the study pertaining to input data, the color green—techniques and algorithms, and the color orange—output data.

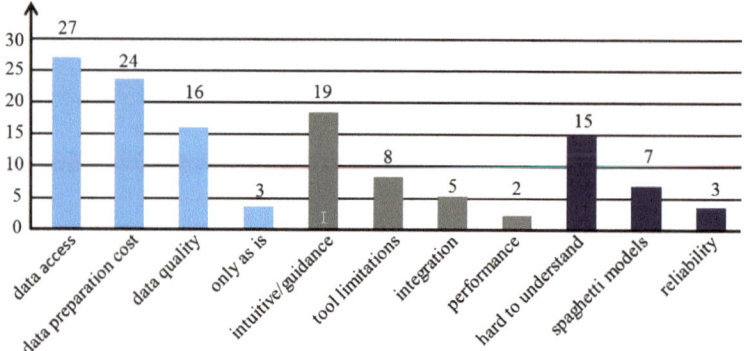

Fig. 1.7 The most commonly noticed flaws of process mining tools and methodologies. *Source* Author's own elaboration on the basis of Claes and Poels [73, pp. 1–5]

Problems pertaining to the use of process mining can be divided into three main groups [71, 73, 74].

1. Technical problems

   1.1. Lack of standardization, which is often accompanied by the poor quality of IT systems event logs. Data may be saved in records with an unknown structure or with an insufficient enough structure to enable process analysis. They may also have different identifiers in different systems or there may be no possibility of defining constant relations between data from different systems, etc. Process Mining Manifesto describes maturity levels for event logs [71, p. 7];

   1.2. No possibility of the integration of analysis of data from sources which are not integrated with or not included in a uniform ICT architecture.

2. Algorithmic problems

   2.1. Problems resulting from the necessity to balance between qualitative criteria, such as conformance, simplicity (transparency), precision, and generalization. This is especially visible when we consider the fact that an event log may include unusual data which might constitute informational noise, or records of a crucial process individualization, which generates new knowledge.

3. Organizational and cultural problems

   3.1. Poor organization of systems data entry. A particularly crucial problem in this regard is entering multiple data not during process performance itself, but upon its completion (e.g. entering all therapeutic data during patient dismissal from the hospital);

   3.2. Leaving out crucial data or entering false data, which are the result of poor work culture or cultural barriers, manifesting themselves in the reluctance to share knowledge uncovered in the course of performing work: viewing process mining as an adversarial system of personnel invigilation, which one must resist;

3.3. Bottom-up discovery of detailed processes without tying the processes to a broad-based outlook on the organization through e.g. the process architecture, which is tied to the risk of adverse sub-optimization or focus on the optimization of non-essential or even unnecessary processes.

Nevertheless, despite the above-mentioned challenges and problems, process mining is at present one of the most promising methodologies, which in connection with other types of analyses allows for the uncovering and dissemination of knowledge on process performance. Its main advantages rest in the objectivity, speed, and flexibility of uncovering new knowledge or presenting the results of verifying old knowledge [67, 73, p. 4]. Of course, the same can be achieved in the course of a traditional audit or some other statistical analysis of processes, but in such circumstance the duration and the cost of execution will be much longer and higher, and will generate averaged results, which will not account for each specific process performance—sometimes numbering in the hundreds. For this reason, process mining tools and methodologies will see increasing use. After all, the goal is to waste no time in finding examples of performances which have led to beneficial (much better than the rest) or negative results—and share this knowledge throughout the organization at a faster pace than the competition.

1.4.4 Robotic Process Automation

Robotic process automation (RPA) is a software class which allows for the automation of routine, usually mass-repeated business processes or tasks, which are usually supported by several IT systems [75, 76, p. 3]. This software class has been developed to respond to the overlapping of multiple factors, among which are the following:

- the complex nature, long execution times, and high costs of projects pertaining to application integration
- the complex nature, high costs, and essential risks of project pertaining to the modification of traditional legacy systems
- the personnel's use of applications and data provided by business partners, the administration, or simply available online. This often prevents the modification or integration of applications which provide or acquire data to and from the organization, albeit are not the property of the organization itself. In effect, this prevents the organization from introducing modifications or familiarizing itself with the structures of data with a view to the application's integration with existing systems.
- lack of a process-based nature of ERP systems, which support functional fields, but not the processes performed in the organization.

Such factors result in employees performing business processes having to make use of several applications, manually entering or copying data between them. RPA tools have the goal of reducing employee workload by delegating such repeated, routine tasks to a software robot—a dedicated computer program [77]. In effect, this a method of work automation, the realization of a principle known since ancient times, in which monotonous and routine human work, or work which surpasses human capabilities in terms of strength, has first been delegated to animals, then to steam engines, electrical devices, and in current years—to industrial automation and IT. For this reason, robotic process automation, is not in and of itself a new concept or even a management methodology, as much as a tool enabling organizations to quickly achieve such benefits as:

- shortening delivery times
- lowering total costs
- minimizing the number of mistakes [78].

RPA tools automate data exchange and allow for the synchronization of selected functions offered by multiple applications without the need of accessing or modifying their source code. The implementation of robotic process automation should in principle offer quick benefits, as it is not usually tied to the necessity of introducing modifications to existing IT systems, nor to the necessity of reengineering the organization's processes. Solutions offered by RPA often make use of screen scraping. This is a technique, thanks to which a computer program—a screen scraper—collects output data from another program by way of reading data presented on screen and initially intended for a human user instead of another program. In effect, it is a substitute for program integration, which allows for the de facto flow of data without the necessity of preparing an integrating interface. When compared with traditional solutions, it allows for saving both time and resources as early as in the preparation and implementation stage [76, p. 5], as it operates on the level of an existing user interface. However, the main time and cost savings of an RPA solution are generated thanks to substituting human work with a program.

It seems possible to implement RPA quickly, as it does not require the rationalization or optimization of business processes, nor personnel training. However, it does require knowledge of the algorithm of process performance prior to performance itself. RPA tools perform a business process in accordance with a built-in algorithm decidedly quicker than an employee—they do not get sick nor require time off, etc. Well-configured software robots always work quickly and without mistakes 24/7/365, provided that they do not face an unforeseen deviation from the implemented algorithm. The emergence of such a situation may cause them to pause work, work illogically, or force the organization to come back to traditional human-based process performance. At the same time, even a small change in the process, which would require the personnel to read an updated instruction manual, listen to a several-minute instruction, or ask a co-worker about the changes in question, in the case of RPA tools requires time-consuming, and, in effect, costly reprogramming.[6]

In effect, the use of RPA tools is possible for processes, their fragments, or single tasks or cases, whose data structure "algorithm" of performance, including rule-based decision making, is unchangeable and known to the last detail or changes in time periods which are much longer than the duration of performance itself [80]. In accordance with the process qualification presented in Fig. 1.1, it is possible to use RPA in the case of:

- static processes
- fragments of static processes with ad hoc exceptions
- static fragments of unstructured processes with predefined fragments

Unfortunately, due to the present construction of existing RPA tools, their use cannot pertain to dynamically managed business processes. However, it may concern their static fragments or particular tasks. As Fig. 1.8 demonstrates, vendors of IT systems already enable their users to resort to RPA technologies in the course of process performance in a way which is tailored to the nature of specific tasks or cases (autonomic or hybrid).

The problems which arise the most often in the implementation an maintenance or RPA tools are [81–84]:

1. Lack of flexibility of process execution;
2. The problem of complexity—even at first glance simple processes may contain several decisions or a dozen or several dozen different variations, which all require programming;

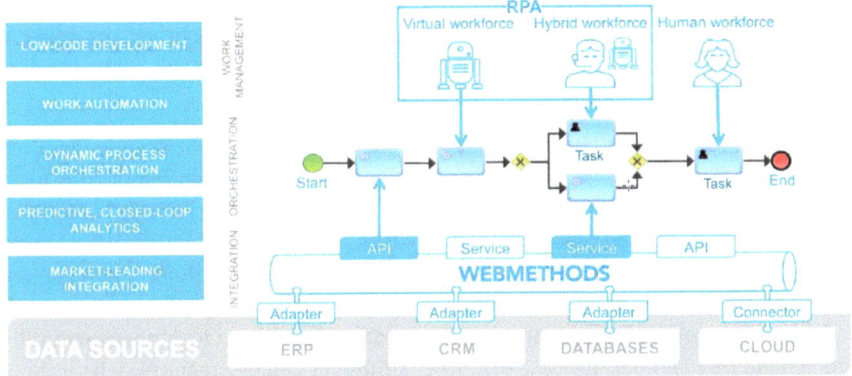

Fig. 1.8 Dynamic apps platform—integrated process, case and robotic automation. *Source* Soft-wareAG [76, p. 12]

---

[6]This is not a new problem connected to RPA tools alone. Davenport and Prusak [79, p. 15] provided the example of a company which decided against using robots by stating: "Using robots was good, but now we're discovering that using people is actually faster".

3. Problems with the use of unstructured data (e.g. scans or photographs, audio or video files). In the human-performed processes, interpretation and decision-making on the basis of such data are usually unproblematic. In the case of software robots, however, the interpretation of unstructured data requires detailed preparation each time and is nevertheless prone to mistakes;

4. Problems tied to deviations or exceptional situations (e.g. wrong values or lack of data required to make a decision), requiring prior analysis and servicing;

5. Problems with making wrong decisions due to changes in the context of process performance which are unseen by the software robot (e.g. making a payment on time, albeit to an account of a bankrupting vendor, who is sure not to supply the ordered products or services). Failing to notice signals from the organization's environment may result in longer, costly reaction times in regard to ongoing changes;

6. Problems with automating the wrong processes (e.g. unstructured) or forcing automation of entire processes, even if they are e.g. static processes with ad hoc exceptions, which cannot be automated with the use of RPA. Their automation may result in "killing their intelligence" by reducing dynamic processes to several simplified patterns, the result of which may be deriving them of the effective possibility of reaching their goals;

7. Communication problems between IT and business in a situation in which IT is often inexperienced with RPA tools and business has far-reaching expectations drummed up by RPA vendors, stemming from its quick implementation;

8. Ethical threats and threats to safety deriving from responsibility for the decision-making process, when the work of robots may be perceived and interpreted as the work of human (e.g. in applications pertaining to healthcare or finance);

9. IT and organizational problems with maintaining an environment in which RPA programs emerge beside existing systems;

10. Limiting the implementation of RPA to automating existing processes, without the analysis of their necessity and efficiency, which may result in the benefits considerably failing to meet expectations.

Even a preliminary analysis of the above list demonstrates that despite the often exaggerated promises of vendors and consultants, the implementation of RPA tools with a very narrow group of processes in mind, while fulfilling rigorous conditions, may be quick, cheap, and risk-free. In most cases, like every new and emerging technology RPA requires cautious calculation of the benefits and careful implementation on a well-selected group of processes.

Process mining techniques and robotic process automation techniques naturally supplement one another. Process discovery or model enhancement may be used to teach processes to software robots on the basis of actual process performance, and not by way of designing processes from scratch by personnel teams. Conformance checking may be used in the course of performance by RPA to check for deviations, anticipate problems, and signal the emergence of situations which require the software robot to handle the performance of a given process to human workers [85].[7]

1.4.5 Machine Learning

Intelligence is the capacity to learn and manage in new situations [86]. The term *artificial intelligence* (AI) has two meanings:

- First, it designates a research field common to both IT and robotics, the research goal of which is the development of systems performing tasks which require the use of intelligence when performed by humans;
- Second, it designates a feature of systems, which allows them to complete tasks which require the use of intelligence when performed by humans [87, pp. 235–236].

In both instances, the focus is on the ability to act purposefully in previously unknown situations. This, in turn, means that the domain of artificial intelligence definitely does not rest in traditional, static business processes. In their case, the people or software robots performing the processes are dealing with known, routinely repeated situations. It is fruitless to perform pattern recognition on a process which is always performed in the same way, though of course it is possible to acquire and analyze information on the broader context of performance, which may have an influence on its ex post innovation.

The most common fields of research and practical application pertaining to artificial intelligence are:

- speech recognition;
- image recognition;
- machine learning (also deep learning);
- expert systems;
- neural networks;
- technologies based on fuzzy logic.

From the perspective of process management, at present the most important field of research and practical application within the scope of artificial intelligence is machine learning (ML). Its goal is to enable programs to prepare forecasts (predicting) and make decisions on the basis of constructed models, as well as learn how to problem solve in a way analogous to the cognitive functions of the human brain [87, pp. 230–231]. This term also encompasses the management and analysis of *Big Data*, which forms the basis of formulating predictions or recommending decisions. The actions of a learning machine may be autonomous or focused on supporting the

---

[7]"For example, RPA vendor UIPATH and process mining vendor Celonis collaborate to automatically visualize and select processes with the highest automation potential, and subsequently, build, test, and deploy RPA agents driven by the discovered process models." [85].

actions of humans cooperating with the program. In accordance with Gartner's [88] recommendations, even now, in the early stage of the development of AI, human support by artificial intelligence is currently the best method of raising the efficiency of process performance.[8] It will usually encompass the processing and analysis of data with a view to:

- preparing multiple variants of decision proposals;
- pattern recognition;
- identification of deviations and analysis or prediction of their results.

One sample application of machine learning is the analysis of *Big Data* acquired thanks to the Internet of Things. In order to recognize patterns of behavior in consumers or analyze results of process mining in regard to therapeutic processes in the scope of conformance checking with effective clinical pathways, or enhancing clinical pathway models. The performance of such tasks would require a long time and would be tiresome, if it were performed by a human working with source data. However, it is much more effective and may be even performed online when the role of the human is to make a decisions on the basis of proposals which are prepared and updated by an artificial intelligence in an ongoing manner. This would enable the employees using (or perhaps "cooperating with") an AI to focus on the more creative and innovative aspects of process performance instead of the multivariate, tiresome, routine calculations. Furthermore, the use of artificial intelligence allows us to analyze data encompassing a much broader context of the process than humanely possible. Such an approach is justified because in our present understanding, artificial intelligence does not come from some esoteric, genius algorithms, but from the use of relatively simple techniques on massive quantities of data, which are not necessarily well structured. For this reason, AI, ML, and data analysis should be connected with *Big Data*, which in this case may also encompass unstructured data (e.g. image or audio files) [89].

Constant process improvement with the use of artificial intelligence will let organizations arrive at process models, metrics forecasts, risk assessments, and recommendations for multivariate decisions which are increasingly closer to reality. For this reason, it should not come as a surprise that multiple BPM vendors focus on artificial intelligence as a method of optimizing processes and reducing their complexity from the perspective of process performers. Thanks to the application of e.g. machine learning, artificial intelligence allows us to make another qualitative step in process management through the online use of data on the entire context of the performed process and the entire available (or acquired in the course of a simulation) knowledge of the organization. To this end, it is essential to combine all of the aforementioned technologies with BPMS systems into an all-encompassing IT environment supporting process management. Such an environment would allow us

---

[8]A more serious topic, which exceeds the scope of this work, is the shape of the cooperation of the humans with artificial intelligence when it will achieve actual intelligence, that is, when it will be capable of learning and solving problems unknown thus far in an unlimited capacity (e.g. From learning to tie one's show to formulating the theory of everything). The author is convinced that then mankind will be faced with much more serious problems than the efficiency of business processes.

to combine the quick-paced nature of RPA with the analytical possibilities of process mining and the flexibility of iBPMS systems including the functionality of case management with a view to increasing productivity, lowering costs, and making work easier [90, 91].

1.4.6 Search for New Directions of the Development of Process Management by Systems Vendors

Despite having reached the end of the cognitive potential of traditional process management without the appearance of a new, cohesive, theoretically sound concept which would be able to respond to the needs of the organization, new methodologies and tools pertaining to process management are still being developed in practice [92, p. 2]. Both process management and BPMS, as well as case management and CMS are evolving rapidly in response to the pressures exerted by the client [93]. In contrast to theoreticians, creators and suppliers of implementation methodologies and BPM had to take into consideration the fact that only 30% of processes within organizations are static in nature, and that due to changes to the work culture and new technologies this percentage is becoming even lower. In order to meet client requirements, multiple companies which for years have specialized in BPM solutions began to create solutions in the form of hybrids of BPMS and CMS (Appian, BPM'online, Bizflow, IBM, K2, Kofax, Pegasystems, PNMsoft) in turn, proponents of case management had to account for the requirements of the users in terms of preparing and approving specific scenarios of performance, which are sometimes unofficial or have the form of suggestions (de facto predefined patterns or process scenarios). At the same time, the pressures of the users of Case Management System (CMS) tools have first led to the adoption of business rules, then process patterns, and finally process scenarios in the form of business process models prepared in different notations [94, 95]. For example, the Adaptive Case Management system offered by ISIS Papyrus allows for the modeling of processes in BPMN, EPC, and UML notations. At the same time, on March 12, 2015 Gartner published the first *Magic Quadrant for BPM-Platform-Based Case Management Frameworks* report [96], and less than a week later another *Magic Quadrant For Intelligent Process Management Suites* report [97] was published, which was partly prepared by the same analysts and for the most part covered the same range of products. Furthermore, 2015 saw the publication of Forrester's and Gartner's reports on traditional BPM Suites and Case Management Systems, as well as similar fields of operation, document management systems (DMS), and workflow systems [98, 99]. 2016 was no different. *The Magic Quadrant for Intelligent Business Process Management Suites* was published on August 18, 2016 [100], while the *Magic Quadrant for BPM-Platform-Based Case Management Frameworks*—on October 24, 2016 [101] and The Forrester Wave™: Dynamic Case Management Q1 2016 report on February 2, 2016 [102]. In 2017, Gartner did not publish another version of the BPM-Platform-Based Case

Management Frameworks report. Instead, Gartner included case management functionality in the requirements for Intelligent Business Process Management Systems (iBPMS) [103]. Case management and the possibility of introducing changes to the principles of operation for applications, including process models, by the users themselves, are according to Gartner an obligatory element, the absence of which disqualifies a BPMS system on the market in 2017. This multitude of analyses and reports on changes offered on the market of methodologies and IT systems, as well as their evolution, shows just how intensely business process management is evolving, and how we are missing an in-depth reflection as regards management theory; one which would attempt to unify and provide direction to the numerous business experiences and the development of BPMS/CMS systems [36, p. 60, 104].[9]

1.5 The 4th Wave of Process Management—"Business Process and Knowledge Management"

As has been demonstrated in Table 1.1 of Sect. 1.2 to this chapter, during the hundred-year-old development of process management, each subsequent stage of development had been the result of pressure exerted by external factors. In response, in each stage of development process management has not been reinvented from scratch, but it has been supplemented with new elements, which allowed for its extension. As has been discussed in Sect. 1.3, the direction of development of process management is at present determined by two mutually stimulating and strengthening global trends:

- the development of information and communication technologies (ICT),
- changes in business culture and social conventions.

As a result thereof, organizations are now facing surmounting challenges in terms of:

- further changes to social norms due to widespread digitization of work and life (forced digitization);
- the growing digitization of business;
- required individualization of processes with the use of the knowledge of the employees and new technological possibilities (like those offered by machine learning and artificial intelligence).

In the context of all of the above-mentioned challenges, we can in fact see the influence of:

- globalization;
- social and mobile technologies;
- *Big Data;*

---

[9]"After 20 years of debating BPM theory, the achievable business benefits of BPMS implementations remain based on vendor sponsored anecdotal evidence and lack long-term, scientific validation!" [29, p. 60, 36].

- the Internet of Things (IoT);
- automation and robotics, including robotic process automation (RPA) [105];
- implementing elements of machine learning and artificial intelligence (AI) [87].

Facing these challenges in the knowledge economy, which is characterized by the fast pace of information flow and operations, as well as constant change, requires organizations to maintain efficiency and renew their competitive advantage, which is fragile in the hypercompetitive environment [106, 107, pp. 56–60]. As studies held since over a decade demonstrate, the use of the organization's intellectual capital has a large influence on both of these goals [108, 109, pp. 1–4], which has been increasing in the last years. Studies on the relation between tangible and intangible capital in organizations show that this ratio amounted to 30:70 in 1929, and in the year 1990—to as much as 63:37 [110, p. 57]. Research held in German enterprises in the years 2010 and 2014 demonstrates that the present influence of intellectual capital on business success has been evaluated by the respondents as an average of 7 on a ten-point scale, while the influence of tangible capital has been evaluated as a 5.2, or over 25% lower [109, p. 3]. The studies also analyzed the methods of operation and the use of intellectual capital. Studies held by Kianto et al. [111, pp. 1473–1482] have shown that intellectual capital is rather static in nature. Its use as a value-creating solution requires the creation of an environment within the organization, in which this capital will be tied to dynamic knowledge management.[10] Only then will process-driven knowledge have a significant influence on the creation of added value.

The development of quality assurance systems maintained by the International Organization for Standardization (ISO) is going in a similar direction. The new ISO 9001:2015 standard includes for the first time the requirement of viewing knowledge as the main resource of the organization, which warrants systematic management. It also introduces, or perhaps extends the Deming cycle with information on the understanding of the methods and goals of managing knowledge in its subsequent stages. At the same time, the goal of the updated standard is the more rigorous approach to efficient knowledge management. According to the authors of the norm, a process-based orientation provides a solid foundation for the mutual connection of quality and knowledge management.

This leads to the natural bringing together, or perhaps even the unification, of process management and knowledge management [112]. The integration of both these concepts on the foundation of acknowledging the empowered role of the employees and the new role of the management enables the creation of a concept which does not have the shortcomings of traditional business process management signaled in Sect. 1.3 of this chapter. This, in turn, allows for the restoration in process management of the constant, systemic possibility to respond to fast-paced, qualitative, unforeseen changes in the hypercompetitive environment. In the knowledge economy, process management cannot be limited to the routine, repeated performance of the same—even the best optimized and managed—process [25]. Due to the nature of business, it must also encompass unstructured process, which require

---

[10]"(…) we have discussed these from the static «asset» perspective of IC stocks and the dynamic «process» perspective of KM practices" [111, 1480].

real-time knowledge management and the performance of which is dependent on the knowledge (including the experience) of the knowledge worker and the individual requirements of the client (the context of performance). Process management must be dynamic in accordance with the definition presented in the beginning of this chapter in order to allow knowledge workers to use their knowledge (including tacit knowledge) in practice in all performed business processes. The goal, which has been postulated by Drucker [113], is to raise the efficiency of the work of the knowledge workers in a way which is analogous to raising the productivity of manual laborers in the 20th century.[11]

As studies shows in Table 1.2, organizations are forced to manage business processes and knowledge in a uniform fashion—in order to create conditions for the efficient use and the development of the organizations' intellectual capitals. This requires process management to make another qualitative leap in development, which according to the author undoubtedly deserves the name 4th wave of the development of process management. Its main assumption is extending process management to allow for the broadest possible use of intellectual capital as the source of competitive advantage thanks to:

- allowing for the ongoing, widespread use and creation of knowledge by the entire personnel of the organization in the course of business process performance;
- managing processes not through data and information, but also thanks to the organization's knowledge;
- rapid adaptation thanks to eliminating the gap between business needs and IT solutions.

Analogous to previous waves of the development of business process management, in the case of the 4th wave we are able to define the factors which trigger it and the principles it proposes, which supplement (extend) the existing principles of process management.

In the 4th wave of process management, processes are treated not as "imposing a certain course of action," but first and foremost as the source of the organization's current knowledge, which to a considerable degree drives the organization's competitive advantage. The task of process management is not only allowing the organization to perform efficiently today, but also the rapid adaptation to changing conditions thanks to eliminating the gap between business needs and IT solutions. As has been described in subchapter 3 of this chapter, analogous to the previous waves of process management, the 4th wave first emerges in practical solutions which embrace the new possibilities offered by ICT systems.

---

[11] "The most important, and indeed the truly unique, contribution of management in the 20th Century was the 50-fold increase in the productivity of the manual worker in manufacturing. The most important contribution management needs to make in the 21st century is similarly to increase the productivity of knowledge work and the knowledge worker." [113].

Table 1.2 Fundamental change factors and changes in approach to process management with the inclusion of the 4th wave of process management

Wave of process management		Rules (assumptions)	Fundamental change factors
I	Industrial engineering (1911–1980)	• No process changes or slow pace of process changes • Elimination of redundant actions and unnecessary losses • Division of the process into simple elements • Full expendability of workers performing simple tasks	• Larger product and service variability, which necessitates larger production process variability • Growing significance of intellectual work • Growing focus on services
II	Value chain management (1985–2003)	• Each task or group of task must provide value for the client • The value is dependent not only on the quality of the work performed in the course of different actions or their groups, but also on their coordination as well • Processes within the organization are innovated upon through evolutionary or revolutionary means	• Globalization • Growing volatility and pace of operations • Rapid development of common ICT technologies
III	Evolutionary adaptation to the needs of the clients (2003–2017)	• Process management as a cohesive and flexible system of operation and innovation within the organization • The entire process is being managed from the point of view of the client, margins while also taking into consideration the organization's suppliers and partners • Harmonious use of information and communication technologies (ICT) in order to raise the efficiency of management and shorten the process optimization loop	• Changes to social culture due to the common digitization of work and life (forced digitization) • Growing digitization of business • Required individualization of processes with the use of *BigData* techniques and Artificial Intelligence • Growing importance of knowledge and the practical use of intellectual capital for the organization

(continued)

Table 1.2 (continued)

Wave of process management	Rules (assumptions)	Fundamental change factors
IV Business process and knowledge management (2017–)	• Allowing for the ongoing, widespread use and creation of knowledge by the entire personnel of the organization in the course of business process performance • Managing processes not through data and information, but also thanks to the organization's knowledge • Rapid adaptation thanks to eliminating the gap between business needs and IT solutions	

Source Author's own elaboration

1.6 Conclusions

In the course of over 100 years of the development of business process management the scope of its use in organizations had undergone changes. In the course of the 1st wave of process management, which was initiated by the works of Taylor's [13], it was limited to production processes alone. At present, process management is used in all fields of management, both within organizations and between the organization and its surroundings. The approach to process management had been changing as well. At first, the aim of process management was the formulation and execution of an optimal, ideal method of work. Later, this goal has been supplemented with the side goal of constant process improvement. Later still, another goal of process management was to adapt processes to the changing needs of clients and the rules of competition [23, 114, pp. 96–106]. The emerging 4th wave of process management is tied to the necessity of accepting and using on a day-to-day basis not just two, as before, but thee different approaches to changes to business processes:

- optimizing processes to changes which are evolutionary in nature, usually quantitative or related to the speed of operations. Such changes are usually predictable and their duration is much longer than the duration of performing the business process they relate to;
- adapting processes to changes which are usually qualitative and unpredictable in nature. Such changes are usually the result of abrupt changes in technology or the organization's surroundings;
- individualizing processes which are usually unpredictable and one-off in nature, combining adaptive changes with the need to account for the requirements (and often the limitations) of specific beneficiaries of the business process. They usually require the instant, multifaceted changes to performed business processes, which are dependent on the context of the specific performance.

As has been demonstrated, changes to the concept of process management were introduced in reaction to changes in the conditions of holding business itself. Theoretical reflection supported practical attempts at adapting process management to the changing expectations of the organization. However, the direction of its further development has been determined by practical considerations. Even today, combinations of e.g. process mining and, robotic process automation, and machine learning, which allow for the online use of knowledge with a view to changing process performance in accordance with the individual context of performance, forces organizations to discard in practice the principles of traditional process management. It is now apparent that the current direction of the development of process management is determined by two mutually stimulating and strengthening factors with large dynamics:

- the development of information and communication technologies (ICT)
- changes in business culture and social conventions

The technical possibilities created in effect of the aforementioned factors, as well as the evolution of client needs, lead the evolution of process management in a direction of maintaining the efficiency of operations and the quality of the supplies products and services, as well as the direction of the ongoing, systemic adaptation of operations to the changing needs of the clients and the conditions of process performance. In the case of the 4th wave of process management, the aim is not just "operational perfection", but also the ability to use the entire intellectual capital of the organization in order to be able to offer individualized products and services at the time and place required by the client [109]. This goal falls outside the scope of traditional process management. There is no doubt that the question posted in the introduction—*Is traditional business process management, which distances the decision on the method of performing work from the act of performance itself, capable of being used in the knowledge economy?*—must be answered in the negative. The excessive duration and length of the feedback loop which adapts processes to the changing conditions of performance, the use of the knowledge and experience of a mere fraction of the employees deciding on the course of the performed processes, and first and foremost the dynamic, often unpredictable nature of the majority—70 to 80%—of the processes within the organization are factors behind the need to extend process management to meet the requirements of business in the knowledge economy.

The achievement of this goal requires the use of dynamic business process management, in which process performers are authorized, in the scope of process performance itself, to use their knowledge with the aim of adapting the performed processes to the requirements of the clients and the context of performance. This knowledge is revealed and managed within the organization in a systemic manner, which allows the organization to broaden the scope and intensify the use of its intellectual capital to an extent which is not available with the use of traditional process management. Only then can process management extend not to 20–30%, but to all of the processes in the organization in the knowledge economy. And this is the main factor triggering the emergence of the 4th wave of process management. As experiences from recent years show, failing to acknowledge or negating the changes will not stall the development of practical process management tools and methodologies [115].

References

1. Płoszajski P (2004) Organizacja przyszłości: przerażony kameleon. W kierunku nowej filozofii zarządzania. Retrieved from http://www.allternet.most.org.pl/SOD/Heterarchia% 20prof._Ploszajski_-_Organizacja_przyszlosci.pdf [22.04.2017]
2. Kisielnicki J, Szyjewski Z (2004) Przedsiębiorstwo przyszłości w warunkach nowej ekonomii. Retrieved from http://zti.com.pl/instytut/pp/referaty/ref13_full.html [20.03.2004]
3. Champy J (2002) X-engineering the corporation. Reinventing your business in the digital age. Warner Books, New York
4. Abell D (1993) Managing with dual strategies. Mastering the present. Preempting the future. The Free Press, New York
5. Toffler A (1990) Power shift. Knowledge, wealth, and violence at the edge of the 21st century. Bantam Books

6. D'Aveni R (1995 Aug) Coping with hypercompetition: utilizing the new 7S's framework. Acad Manag Exec 9(3):45–57
7. Murray P, Myers A (1997) The facts about knowledge. Special report, 11.1997
8. Hamel G, Valikangas L (2003) The Quest for Resilience. Harv Bus Rev (September 2003)
9. Onken MH (2003) Temporal elements of organizational culture and impact on firm performance. Retrieved from http://www.emeraldinsight.com/pdfs/200064.pdf [19.05.2003]
10. Senge P (1994) The fifth discipline. The art and practice of the learning organization. Currency Doubleday, New York
11. Smith H, Fingar P (2002) Business process management—the third wave. Meghan-Kiffer Press
12. Drucker P (2000) Zarządzanie w XXI wieku (Managements challenges for the 21st century). Muza S.A., Warszawa
13. Taylor FW (1911) The principles of scientific management. Harper & Brothers, New York
14. Hammer M (1999) Reinżynieria i jej następstwa – jak organizacje skoncentrowane na procesach zmieniają naszą pracę i nasze życie (*Beyond reengineering. how the process-centered organization is changing our work and our lives*). Wydawnictwo Naukowe PWN SA, Warszawa
15. Aalst W (2013) Business Process Management: A Comprehensive Survey. ISRN Software Engineering 2013, Article ID 507984, pp.1-37.
16. Nevins A, Hill F (1957) Ford: expansion and challenge. 1915–1933. Charles Scribners' Sons, New York
17. Porter M (1985) Competitive advantage. Free Press, New York
18. Deming WE (1986) Out of the crisis. Massachusetts Institute of Technology, Center for Advanced Engineering Study, Cambridge
19. Davenport T, Short J (1990) The new industrial engineering: information technology and business process redesign. Sloan Manag Rev 31(4):11–27
20. Hammer M (1990) Reengineering work: don't automate, obliterate. Harv Bus Rev 104–112
21. Hammer M, Champy J (1993) Reengineering the corporation: a manifesto for business revolution. Collins Business Essentials, New York
22. Davenport T (1995) The fad that forgot people. Fast Co Mag 1. Retrieved from https://www.fastcompany.com/26310/fad-forgot-people [12.03.2017]
23. Smith H, Fingar P (2003) Business processes: from reengineering to management. Retrieved from http://www.peterfingar.com/Darwin-BPR-to-BPM.pdf [11.12.2016]
24. Davenport T (1996) Some principles of knowledge management. Strategy Bus 1(2):34–40. Retrieved from https://www.strategy-business.com/article/8776?gko=f91a7 [10.11.2017]
25. Fingar P (2007) The greatest innovation since BPM. Retrieved from http://www.bptrends.com/publicationfiles/SIX-03-07-COL-TheGreatestInnovationSinceBPM-Fingar-Final.pdf [2.12.2017]
26. Gartner IT Glossary (2016) Business process management BPM. Retrieved from http://www.gartner.com/it-glossary/business-process-management-bpm [3.03.2016].
27. Pande P, Neuman R, Cavanagh R (2003) Six Sigma. K.E. Liber S.C, Warszawa
28. Business Dictionary (2017) Hidden factory. Retrieved from http://www.businessdictionary.com/definition/hidden-factory.html [31.03.2017]. ("Activities that reduce the quality or efficiency of a manufacturing operation or business process, but are not initially known to managers or others seeking to improve the process") [4.09.2017]
29. Pucher M (2012) The strategic business benefits of adaptive case management. In: Fischer L (ed) How knowledge workers get things done. Real-world adaptive case management. Future Strategies Inc., Lighthouse Point, Florida, USA
30. Röglinger M, Pöppelbuß J, Becker J (2012) Maturity models in business process management. Bus Process Manag J 18. Retrieved from http://www.fim-rc.de/Paperbibliothek/Veroeffentlicht/352/wi-352.pdf [29.03.2017]
31. Kania K (2013) Doskonalenie zarządzania procesami biznesowymi w organizacji z wykorzystaniem modeli dojrzałości i technologii informacyjno-komunikacyjnych. Wydawnictwo Uniwersytetu Ekonomicznego w Katowicach, Katowice

32. Polish Air Navigation Services Agency (2009) Emergency checklists. Ver. 1. Warszawa: Wieżowa 8 [27.05.2009]
33. Belaychuk A (2011) ACM: paradigm or feature? Retrieved from http://mainthing.ru/item/401/ [21.01.2011]
34. Kemsley S (2010) Runtime collaboration and dynamic modeling in BPM: allowing the business to shape its own processes on the fly. Cut IT J 23(2):35–39
35. Tran T, Weiss E, Ruhsam C, Czepa C, Tran H, Zdun U (2018) Enabling flexibility of business processes using compliance rules: the case of mobiliar. In: vom Brocke J, Mendling J (eds) Business process management cases. Springer International Publishing AG
36. Pucher M (2012) How to link BPM governance and social collaboration through an adaptive paradigm. In: Fischer L (ed) Social BPM: work, planning and collaboration under the impact of social technology (pp 57–76). Future Strategies Inc., Lighthouse Point
37. Deming WE (1982) Quality, productivity and competitive position. Massachusetts Institute of Technology, Center for Advanced Engineering Study, Cambridge
38. Rummler G, Brache A (2000) Podnoszenie efektywności organizacji (Improving Performance). PWE, Warszawa
39. Johnson JC, Manyika JM, Yee LA (2005) The next revolution in interactions. McKinsey Q, pp 20–33. Retrieved from https://www.mckinsey.com/business-functions/organization/our-insights/the-next-revolution-in-interactions [2.07.2018]
40. Kemsley S (2009) Hidden costs of unstructured processes #GartnerBPM. Retrieved from http://column2.com/2009/10/hidden-costs-of-unstructured-processes-gartnerbpm/ [3.04.2016]
41. Austin T (2010) Gartner says the world of work will witness 10 changes during the next 10 years. Retrieved from https://www.gartner.com/newsroom/id/1416513 [2.07.2018]
42. Pucher M (2010) Gartner group 2020: the de-routinization of work. Retrieved from http://isismjpucher.wordpress.com/2010/11/12/the-future-of-work/ [5.04.2016]
43. Swenson K (2010) Mastering the unpredictable: how adaptive case management will revolutionize the way that knowledge workers get things done. Meghan-KifferPress, Tampa, USA
44. Ukelson J (2010) Adaptive case management over business process management. Retrieved from http://it.toolbox.com/blogs/lessons-process-management/adaptive-case-management-over-business-process-management-40002 [7.04.2016]
45. Handy Soft (2012) Dynamic BPM—the value of embedding process into dynamic work activities: a comparison between BPM and e-mail. Retrieved from http://www.bizflow.com/system/files/downloads/HandySoft%20-%20Dynamic%20BPM%20White%20Paper_0.pdf [2.04.2016]
46. Olding E, Rozwell C (2015) Expand your BPM horizons by exploring unstructured processes. Gartner Technical Report G00172387, Refreshed: 22 May 2015; Published: 10 Dec 2009
47. vom Brocke J, Zelt S, Schmiedel T (2016) On the role of context in business process management. Int J Inf Manag 36 (3):486–495
48. Di Ciccio C, Marrella A, Russo A (2012) Knowledge-intensive Processes: an overview of contemporary approaches? In: 1st international workshop on knowledge-intensive business processes (KiBP 2012) June the 15th, Rome, Italy. Retrieved from http://ceur-ws.org/Vol-861/KiBP2012_paper_2.pdf [2.04.2016]
49. Kemsley S (2011) The changing nature of work: from structured to unstructured, from controlled to social. In: Lecture Notes in Computer Science Business Process Management, pp 2–2
50. Nowosielski S (2012) Zarządzanie procesami. Retrieved from http://procesy.ue.wroc.pl/uploads/pliki/procesy/wyklady/ZPRnowosielski WYKLAD.pdf [8.08.2017]
51. Leavitt H, Whisler J (1958) Management in the 1990s. Harv Bus Rev 6/1958
52. Obłój K, Zdziarski M (2004) Raport the conference board "Wyzwania stojące przed prezesami firm w 2005 roku". Harv Bus Rev
53. CEO_Challenge (2015) Creating opportunity out of adversity. Building Innovative, People-Driven Organizations. Retrieved from https://www.conference-board.org/retrievefile.cfm?filename=TCB_1570_15_RR_CEO_Challenge3.pdf&type=subsite [1.04.2017]

54. Pucher M (2010) Agile-, AdHoc-, Dynamic-, Social-, or adaptive BPM. Retrieved from https://isismjpucher.wordpress.com/2010/03/30/dynamic-vs-adaptive-bpm/ [18.03.2016]

55. English L (2009) Quality information applied. Wiley Publishing Inc., Indianapolis

56. Aalst W, Weske M, & Grünbauer D (2005) Case handling: a new paradigm for business process support. Data Knowl. Eng. 53(2): 129-162.

57. Pucher M (2010) The difference between ACM and BPM. Retrieved from https://acmisis.wordpress.com/2010/10/01/the-difference-between-acm-and-bpm/ [21.06.2012]

58. Earls A (2011) The evolution of human-centric BPM. Retrieved from http://www.ebizq.net/topics/bpm_technology_implementation/features/13277.html [5.04.2016]

59. Mathur A (2012) BPM and case management. Retrieved from http://www.bpmleader.com/2012/03/20/bpm-and-case-management/ [7.04.2016]

60. Harmon P (2016) Harmon on BPM: business process and case modeling. Retrieved from http://www.bptrends.com/harmon-on-bpm-business-process-and-case-modeling/ [8.04.2016]

61. Garvin D (1998) The processes of organization and management. Sloan Manag Rev. Retrieved from http://sloanreview.mit.edu/article/the-processes-of-organization-and-management/ [15.04.2016]

62. Harmon P (2014) Harmon on BPM: what is case management? Retrieved from http://www.bptrends.com/harmon-on-bpm-what-is-case-management/ [8.04.2016]

63. Swenson K (2010) Comparison: ACM vs. BPM. Retrieved from http://www.xpdl.org/nugen/p/adaptive-case-management/public.htm [20.02.2016]

64. Fisher L (2012) How knowledge workers get things done. Real-world adaptive case management. Future Strategies Inc., Lighthouse Point, FL, USA

65. Swenson K (2015) To model or not to model. Retrieved from https://www.bp-3.com/to-model-or-not-to-model/ [6.07.2017]

66. OMG (2014) Case management model and notation (CMMN) v1.0. Retrieved from http://www.omg.org/spec/CMMN/1.0/PDF [1.08.2017]

67. Aalst W (2016) Process Mining - Data Science in Action, Second Edition. Springer

68. Kerremans M (2008) Automated Business Process Discovery Improves BPM Outcomes. Gartner Technical Report ID G00164422, Published 23 December 2008.

69. Gartner IT Glossary (2017) Automated Business Process Discovery (ABPD). Retrieved from https://www.gartner.com/it-glossary/automated-business-process-discovery-abpd

70. Sinur J (2010) The top 10 BPM technologies. Retrieved from http://blogs.gartner.com/jim_sinur/2010/03/29/the-top-ten-bpm-technologies/#comments [26.01.2011]

71. IEEE Task Force on Process Mining (2012). PROCESS MINING MANIFESTO. Retrieved from http://www.win.tue.nl/ieeetfpm/doku.php?id=shared:process_mining_manifesto [02.04.2016].

72. Aalst W, Dustdar S (2012) Process mining put into context. Internet Comput IEEE 16(1):82–83. Retrieved from http://wwwis.win.tue.nl/~wvdaalst/publications/p662.pdf [3.04.2016]

73. Claes J, Poels G (2012) Process mining and the ProM framework: an exploratory survey. Retrieved from http://www.janclaes.info/papers/PMSurvey/ProMSurveyExtendedReport.pdf [8.07.2017]

74. Mans RS, Aalst W, Vanwersch RJB, Moleman AJ (2013) Process mining in healthcare: data challenges when answering frequently posed questions. In: Lenz R, Miksch S, Peleg M, Reichert M, Riaño D, ten Teije A (eds) Process support and knowledge representation in health care. Lecture notes in computer science, vol 7738, Springer, Berlin, pp 140–153. Retrieved from http://wwwis.win.tue.nl/

75. Sobczak A (2017) Czym jest RPA (robotic process automation)? Retrieved from https://robonomika.pl/czym-jest-rpa-robotic-process-automation [2018-04-03]

76. Software AG (2018) Four reasons why leading companies are implementing RPA

77. Aguirre S, Rodriguez A (2017) Automation of a business process using robotic process automation (RPA): a case study. Appl Comput Sci Eng Commun Comput Inf Sci. https://doi.org/10.1007/978-3-319-66963-2_7

78. Harmon P (2017) Robotic process automation (RPA). Retrieved from https://www.bptrends.com/robotic-process-automation-rpa/ [2018-03-07]
79. Davenport T, Prusak L (1998) Working knowledge—how organisations manage. What they know. Harvard Business School Press, Boston
80. Tornbohm C (2016) Robotic process automation: eight guidelines for effective results. Gartner Report ID: G00309398, Published: 12 Oct 2016, Refreshed: 05 Feb 2018
81. Edlich A, Sohoni V (2017) Burned by the bots: why robotic automation is stumbling. Retrieved from https://www.mckinsey.com/business-functions/digital-mckinsey/our-insights/digital-blog/burned-by-the-bots-why-robotic-automation-is-stumbling [20.05.2018]
82. Raja A (2017) Seven pitfalls of process automation. Retrieved from https://www.bptrends.com/seven-pitfalls-of-process-automation/ [2018-03-09]
83. Sobczak A (2018) Jakie są źródła niepowodzeń przy wdrażaniu RPA? Retrieved from https://robonomika.pl/czy-siedzimy-na-bombie-zegarowej-czyli-jak-duze-czeka-nas-rozczarowanie-po-wdrozeniu-narzedzi [2018-03-09]
84. UIPATH (2017) Why some RPA deployments fail. And what you can do about it. Retrieved from https://www.uipath.com/blog/why-rpa-deployments-fail [20.05.2018]
85. Aalst W, Bichler M, Heinzl A (2018) Robotic process automation. BISE. Retrieved from https://doi.org/10.1007/s12599-018-0542-4 [21.05.2018]
86. Webster Dictionary (2018) Intelligence. Retrieved from https://www.merriam-webster.com/dictionary/intelligence [20.05.2018].
87. Flasiński M (2016) Introduction to artificial intelligence. Springer International, Switzerland
88. Gartner (2018) Gartner top strategic predictions for 2019 and beyond. Retrieved from https://www.gartner.com/smarterwithgartner/gartner-top-strategic-predictions-for-2019-and-beyond/
89. Garimella K (2017) Cognitive BPM business processes awaken! Retrieved from http://www.dbizinstitute.org/resources/articles/cognitive-bpm-business-processes-awaken [22.05.2018]
90. Buccowich B. (2018) What is the difference between RPA and BPM? Retrieved from https://www.laserfiche.com/ecmblog/what-is-the-difference-between-robotic-process-automation-rpa-bpm/ [06.03.2018]
91. Appian (2017) CXO's guide to robotic process automation. Retrieved from https://connect.convedo.com/hubfs/Whitepapers/rpa/ap_CXO_Guide_RPA_R3.pdf?submissionGuid=12a52603-d0ac-437b-856f-eace002e01f5 9.02.2018 [22.05.2018]
92. Plattfaut R (2014) Process-oriented dynamic capabilities. Springer, Berlin
93. Szelągowski M (2013) Geneza dynamicznego zarządzania procesami biznesowymi. Kwartalnik Naukowy Uczelni Vistula 4(38):41−56
94. Keirstead KW (2013) Fixing BPM—out of the frying pan into the fire. Retrieved from http://www.bpmleader.com/2013/07/12/fixing-bpm-%E2%80%93-out-of-the-frying-pan-into-the-fire/ [3.04.2016]
95. ISIS Papyrus (2016) Adaptive processes. Retrieved from http://www.isis-papyrus.com/e15/pages/business-apps/adaptive-case-management/adaptive-process.html [4.04.2016].
96. Gartner (2015) Magic Quadrant for BPM-platform-based case management frameworks. ID: G00262751; Published: 12 March 2015
97. Gartner (2015) Magic Quadrant for Intelligent Business Process Management Suites. ID: G00258612; Published: 18 March 2015
98. Antunes P, Mourão H (2010) Resilient business process management: framework and services. Expert Syst Appl
99. Bider I, Perjons E (2015) Design science in action: developing a modeling technique for eliciting requirements on business process management (BPM) tools. Softw Syst Model 14(3):1159–1188. Springer, Berlin
100. Gartner (2016) Magic Quadrant for Intelligent Business Process Management Suites. ID: G00276892; Published: 18 Aug 2016
101. Gartner (2016) Magic Quadrant for BPM-platform-based case management frameworks. ID: G00276724; Published: 24 Oct 2016

102. Forrester (2016) The Forrester wave™: dynamic case management, Q1 2016. Published: 2 Feb 2016
103. Gartner (2017) Magic Quadrant for intelligent business process management suites. ID: G00315642; Published: 24 Oct 2017
104. Trkman P (2009) The critical success factors of business process management. Int J Inf Manag 30(2):125–134
105. Forrester (2018) The Forrester wave™: robotic process automation, Q2 2018. Published: 26 June 2018
106. D'Aveni R (1994) Hypercompetition managing the dynamics of strategic maneuvering. The Free Press, New York
107. Drejer A (2002) Strategic management and core competencies: theory and application. Quorum Books, Westport
108. Kianto A, Andreeva T, Pavlov Y (2013) The impact of intellectual capital management on company competitiveness and financial performance. Knowl Manag Res Pract 2(11):112–122
109. Kohl H, Orth R, Steinhöfel E (2015) A practical approach to process-oriented knowledge management. Retrieved from https://www.researchgate.net/publication/280231388_A_Practical_Approach_to_Process-Oriented_Knowledge_Management?ev=publicSearchHeader&_sg=F4BV1WfY1cLzGxDRnPsewmy1LIR0etxb4t7NYV0uq4nXX7utZ9wRUm4b4pO6vAoYXLdTCHCUcpPGM3Y [18.07.2017]
110. Fazlagić J (2006) Zarządzanie wiedzą. Szansa na sukces w biznesie. Gnieźnieńska Wyższa Szkoła Humanistyczno-Managerska, Gniezno
111. Kianto A, Ritala P, Spender J, Vanhala M (2014) The interaction of intellectual capital assets and knowledge management practices in organizational value creation. J Intell Cap 3(15):362–375
112. Taylor C (2012) Reunifying knowledge and business process management. Retrieved from http://citeseerx.ist.psu.edu/viewdoc/download?doi=10.1.1.225.9570&rep=rep1&type=pdf [18.07.2017]
113. Drucker P (1999) Knowledge-worker productivity: the biggest challenge. Calif Manag Rev 41(2)
114. Armistead C, Pritchard J, Machin S (1999) Strategic business process management for organisational effectiveness. Long Range Plan 1(32):96–106
115. Toffler A (1980) The third wave. Bantam Books

Chapter 2
Dynamic Business Process Management

Abstract As discussed in Chap. 1, all organizations which hope to prosper in the 21st century must be process-oriented. At the same time, traditional process management does not encompass the majority of processes in modern organizations, as well as prevents the broad use of intellectual capital. Dynamic business process management is an extension of traditional business process management. It does not share its limitations, which hinder or prevent its use in the knowledge economy. Among multiple names of methodologies or proposals of directions of development for process management presented in literature, such as *agile, intelligent, adaptive*, or *human*, the author has selected the term *dynamic* in order to underline that the actual source of all new possibilities offered by dynamic business process management is the dynamism of the knowledge workers themselves. Not just their knowledge, but also their willingness to work is decisive in terms of whether the course of performance will see agile, intelligent adaptations with the aim of tailoring process performance to the context of performance, which stem from the experiences of the process performers themselves. The concept of dynamic business process management does not substitute traditional business process management, as much as it is its extension based on the three principles described in Sect. 2.2 of this chapter. This chapter presents different dimensions of the change, which taken together form an ecosystem enabling the engagement of the entire (or to a larger degree than before) intellectual capital of the enterprise, working and sharing knowledge in the course of process performance and creating value for the client. This ecosystem consists of philosophies, methodologies, and tools supporting management, or the entire BPM.

Keywords Business process management (BPM) · Dynamic business process management (dynamic BPM) · Process lifecycle · Knowledge management

2.1 Introduction

As has been discussed in Chap. 1, each organization which hopes to prosper in the 21st century must shift its focus from traditional organizational hierarchy and instead focus on processes, which at present are viewed to an increasing degree from the

© Springer Nature Switzerland AG 2019   55
M. Szelągowski, *Dynamic Business Process Management in the Knowledge Economy*, Lecture Notes in Networks and Systems 71,
https://doi.org/10.1007/978-3-030-17141-4_2

perspective of clients and business partners [1]. In practice—in accordance with the 3rd wave of development of process management—this shift in focus requires the systemic management of the ongoing adaptation of processes to changing conditions. One spectacular confirmation of this paradigm is quality management, in which a process-centered approach, the necessity to define and maintain knowledge that is essential in the performance of processes, and partnership between all interested parties have assumed the status of normalized operations subject to certification [2]. What, then, is the source of problems that organizations are facing, including even those organizations which implemented process management in accordance with the ISO norm and have been issued the certificate to back it up? Why the countless jokes about procedures and instructions describing each and every aspect of operations, including e.g. the correct method of "plugging electrical plugs into sockets"? Perhaps the issue at hand lies in the desire to fulfill all norms to a painstaking degree, regardless of the changing conditions? The desire for achieving the complete description and the ongoing "optimization through standardization" of the largest possible number of processes—in accordance with the principle that the larger the number of procedures, the better, as procedures free the organization from risk, and the process performers themselves—from the need to evaluate their actions and be held responsible? Nevertheless, such an approach also leads to significant threats:

- the lack of responsibility for the work performed;
- a visible reduction in the innovative character of operations due to constraints imposed by procedures, or—in accordance with the new norm—by business processes;
- the "averaging" of processes within the organization due to the preparation of their models to account for the statistical client—without the option of tailoring processes to specific, individual subjects.

Unfortunately, due to rapid technological changes and changes in the competitive environment the aforementioned threats negate at an increasing pace the benefits which stem from implementing traditional process management. At a time when the competition is merely "a click away," organizations are challenged by the necessity of adapting their fundamental business processes to the requirements of individual clients [3, p. 29; 4]. And because the expectations, habits, and capabilities of the clients are diverse, or even conflicting, the key to success no longer lies in the optimal, "averaged" business process, but in the most skillful dynamic structuring of business processes in accordance with client expectations. Since—as has been demonstrated in Chap. 1—traditional process management does not allow for operating in accordance with the aforementioned requirements, it becomes crucial to ask ourselves whether it is possible to extend process management to meet the demands of the knowledge economy? What are the conditions of such an extension and what are its consequences? What goals should be set for (dynamic) process management in the knowledge economy?

The aim of this chapter is to present the concept of dynamic business process management and its consequences, including the possibility of reunifying process management with case management.

2.2 The Definition of Dynamic Business Process Management

As has been demonstrated in Chap. 1, it is crucial to extend traditional business process management, in which processes are performed in accordance with a predefined standard model, and the process owner monitors the values of performance indicators and introduces periodical changes to the standard process on a predefined timescale. Due to the necessity of maintaining a pace of operations which satisfies client needs, as well as adapting process performance to ongoing changes and specific contexts of performance, it becomes impossible for the process owner or manager to introduce changes on an ongoing basis. In large, medium, and small enterprises it is crucial to empower direct process performers with the capability to dynamically modify processes in the course of performance itself. Only in this way will it be possible to retain the pace and the flexibility of operations, which is especially missing in medium and large organizations. One proposed solution is my own concept of dynamic business process management, in which the term dynamic business process management is defined as follows:

> Dynamic business process management is understood as management which enables organizations to react to fundamental conditions of operation (both internal and external) and cater to the individual needs of the clients in a timely fashion (and in the case of critical factors—practically instantaneously) on the basis of process adaptations entered in real time in the course of performance itself by their direct performers.

Dynamic business process management is an extension of traditional business process management. Alongside the numerous names of methodologies suggested in literature, or alongside the suggestions of the future directions of development of process management, such as *agile, intelligent, adaptive,* or *human,* I have chosen the term **dynamic** in order to stress that the actual source of all further possibilities offered by dynamic BPM is the dynamism[1] of the knowledge workers themselves, which rests in identifying, performing, and analyzing the performances of processes. Both their knowledge and their motivation are crucial in ensuring that agile, intelligent adaptations—stemming from the experience of the performers themselves—will be introduced in the course of process performance itself, with the aim of adapting

---

[1] Dynamism—The quality of being characterized by vigorous activity and progress [5].

the given process to the specific context of performance. Both the knowledge and the dynamism of the knowledge workers is essential if the latter are to attempt to improve the performed processes (in the form of limited, creative experimentation) within the limits of their executive privileges, with a view to creating innovative solutions which allow for the verification of old, as well as the creation of new knowledge [6, pp. 36, 37]. As will be demonstrated later in Chap. 3, knowledge presumes a certain potential to undertake action. However, it is through energy, action, and motivation that this potential is realized in the form of specific actions, in this case—stemming from the three components which comprise knowledge: information, the operational context, and experience.

In accordance with the correspondence principle, the concept of dynamic business process management does not replace traditional (static) business process management as such, as much as it is its expansion, that is, it allows us to understand and describe (accurately anticipate or manage) a broader part of reality than traditional management.[2]

The concept of dynamic business process management does not replace traditional (static) business process management as such, as much as it extends it. In accordance with the principle of the economy of thought [7, pp. 50, 51; 8], this extension has been implemented in accordance with the following three principles.

The 1st Principle of Dynamic BPM: Comprehensiveness and Continuity

The implementation of dynamic BPM should in the least include the standard operational process describing the core activities of the enterprise. Only then is it possible to avoid suboptimization, manifesting itself in one of the organizational units improving its results at the cost of the entire organization [9, pp. 31–41]. For this reason, the identification and subsequent optimization of processes should not just encompass a single organizational unit or a single process (e.g. acquisition or production), but at least all of the fundamental processes within the organization [10]. Because the goal is the optimization of processes from the perspective of the client, it should be undertaken not through the lens of a single organization or enterprise, but with an

---

[2]The correspondence principle—a principle stating that within science, an older theory may be superseded by a newer theory only when the newer theory, beside explaining new facts, also explains facts that have been explained by the older theory. Usually, facts explained by the older theory are in the newer, more general theory considered as borderline (special) cases [7, pp. 50, 51; 8].

One example of applying the correspondence principle is classical mechanics being superseded by the theory of relativism, which expanded the original theory. This does not mean, of course, that classical mechanics is no longer valid. For much smaller velocities than the speed of light (e.g. smaller than 1% of the speed of light, or 10.8 mln km/h) and far from large centers of mass, classical mechanics still describes reality with very high accuracy. At the same time, all of the predictions made by the theory of relativity in regard to velocities which are much lower than the speed of light are completely in sync with the predictions made by classical mechanics. However, thanks to the theory of relativity we are capable of understanding why our predictions in regard to high velocities and large masses were flawed and what changes needed to be incorporated into the model.

By analogy, dynamic business process management identifies factors which limit the use of traditional business process management in the knowledge economy, points to the fields in which it still may be used as a special case within dynamic business process management, and defines more general principles of using (implementing) dynamic BPM.

overlook on a group or a network of organizations producing and supplying goods or services. These might be organizations operating within a single value chain or comprising a virtual or networked organization, etc. In this view, the specific method of cooperation and the capital ties are irrelevant. What is crucial, as in the case of a single organization, is the rejection of barriers dividing the organizations and the optimization of the entire value-making process. From the perspective of the client, the goal is not the minimization of e.g. the costs or the time of work of the organization which is the direct seller or the general contractor. Usually, the client both lacks information on and has no interest in the specific method of dividing costs between the organizations cooperating in the production or supply of products or services. The client is only interested in the delivery date for the product or service and its end cost. For this reason, the goal is to minimize the inclusive costs and total supplies, while simultaneously reducing the total time of execution [11, p. 136]. This approach significantly broadens the range of possible options of raising efficiency, as well as often lowers the time of process performance and project execution by means of optimization which takes into account those operations which fall outside the scope of a single organization (e.g. supplies, warranty service) within the entire value-making process, which defines the end cost for the client [12, pp. 95, 96; 13, p. 152]. At the same time, it raises the level of comport in terms of client cooperation. Thanks to the integration of operations and their focus on a single goal, the client is finally able to deal with a single organization instead of a number of separate units, departments, or subjects working toward their own goals, optimizing their own margins, or repeatedly undertaking e.g. preliminary measures before initializing subsequent stages of work within a single process.

Dynamic business process management—in accordance with the first two points of New 7-S's, or Richard D'Aveni's concept of hypercompetition [14]—allows for the practical use of the fact that management within the organization and cooperation with other organizations are merely two different variations of the same task [15, pp. 37, 38]. Because internal control operations might at present be automated to a significant degree, growing attention should be given to the organization's surroundings, as they are the main source of uncertainty and opportunities for the organization. In consequence, management in the 21st century must be directed outward. The faster the pace of changes, the larger should be the focus of the organization's management on the client and the possible innovations, which enable the maintenance or creation of competitive advantage—and not on internal, routine operations, such as preparing financial reports. In an economy, in which network enterprises and enterprise networks compete with one another, it becomes essential to adapt the practice of management to a situation in which competitive advantage does not only stem from the internal resources at hand, but, to an increasing degree, from relations with external units [16, p. 224].

Another argument in favor of the comprehensive nature of process management within the organization is the necessity of different organizational units to adapt to the needs of the clients at a uniform pace. Given the rapid pace and the qualitative nature of changes in the competitive environment, as well as the cooperative nature of operations, it is no longer possible to administratively upkeep different paces of change in different organizational units. Such an attempt would, in the short run, lead to the dissolution of bonds between subsequent processes which create value for the client, as well as the emergence of numerous problems in terms of workflow and information flow. Within a single organization, it is not possible to e.g. introduce dynamic innovations to management in a sales department, all the while having the rules of production remain the same.

The 2nd Principle of BPM: Process Execution Should Guarantee Evolutionary Flexibility

Employees performing a given process are provided with a "standard process," which is designed in accordance with the best knowledge to date. The process is standard "as of today," as the standards change constantly in response to client needs, technological changes, or new organizational experiences. Because in practice no two conditions of process performance are the same (e.g. two identical consulting projects, construction investments, or diagnostic-therapeutic processes), processes which are standard "as of today" must be dynamically adapted to the specific conditions of a given performance by their direct performers. The PDSA process improvement cycle, which is traditionally used with the participation of the organization's management (also called the Deming cycle after its creator's surname), which consists of process modeling, performance observation, drawing conclusions, and using such-acquired knowledge to improve processes, has proven itself to be too slow and insufficient [17], especially given the fact that in the case of the conflicting expectations of different groups of clients, it might as well turn out that it is impossible to create a standard ("averaged") process, which would be acceptable to of the current clients of the organization.

Processes should be defined and implemented in such a way, as to enable supplements to, or even overhauls of, tasks comprising the process on every step of its performance by direct process performers (e.g. medical doctors, project coordinators, salesmen, or other knowledge workers). It is them, and not, as it had been before, just the process owners, who must be authorized to perform limited experiments in the form of changes introduced in the course of process performance [6, pp. 36, 37]. With the help of a supporting IT solution, they should be able to perform actions, or even entire fundamental processes which are not included in the standard process "as of today." From a practical standpoint, one crucial issue is to define the scope of such executive privileges to introduce changes by different groups of process performers. Too narrow a scope of authorization will limit the dynamic management of processes, reducing it to traditional process management, since process performers will be de facto forced to perform the "standard process as of today." On the other hand, too broad a scope of authorization might result in chaos and disorder within the organization, which should be prevented by creating additional coordinating mechanisms, etc. The scope of authorization to supplement or change the performed process, which defines the field for active experimentation on the part of the knowledge workers, should be dependent on:

- the nature of the process;
- its relation to the process describing the fundamental operations of the organization;
- the knowledge and the level of engagement of the knowledge worker;
- the actual mechanisms of monitoring process performance available in the organization.

By observing the actual multiple performances of the process and their end results, the process owner must be authorized to supplement and remodel the standard process in accordance with the latest available knowledge and with the use of goo business practices, understood as those practices which will lead to success in subsequent iterations of the process. In this context, success can be achieved by preventing mistakes (e.g. supplementing the decision-making process with control tasks and checklists) or by more time-efficient and effective performance, which generates better end results (e.g. by a different division of work, leaving out unnecessary decision points, a better understanding of client needs, faster coordination of work with subcontractors), or perhaps through innovative actions, which were not even on the table when the process was in its planning stages. These are often factors which were known beforehand, albeit ones which were not given much thought, as accounting for all possibilities in the process description was too expensive or impossible due to lack of space.

The 3rd Principle of Dynamic BPM: Processes are Considered Completed Only after Having Been Documented
The implementation of dynamic business process management should be performed in such a way, as to ensure that processes are considered completed only after having been documented in an IT system supporting their performance (e.g. BPMS, CMS, ERP, CRM, HIS, EMR). Only then will it be possible to compare the specific performance of the process with the process definition (the standard process "as of today"). And only then will it be possible to compare both in order to obtain information on all innovations introduced by process performers, as well as their results. In such circumstance, it will be possible to speak not of ex post management, but of ongoing, dynamic management on the basis of information which is constantly reaching the organization's management. The ongoing documentation of process performance allows for the ongoing uncovering of the unwritten rules which are in effect in the organization, which in allows for their critical evaluation and transformation into "written" rules—in the form of process descriptions, procedures, or rules and regulations. First and foremost, this enables us to resolve conflicts between existing, written-down principles and processes and their uncovered, unwritten counterparts. Thanks to the enormous effort of individual or team "heroes,"[3] this prevents e.g. committing to misdesigned processes and bad organization [17; 18, pp. 73, 74]. In such circumstances, thanks to the ongoing monitoring of objective indicators for processes and the process workflow itself, it is possible to quickly correct the process

---

[3]"One gets a good rating for fighting a fire. The result is visible; can be quantified. If you do right the first time, you are invisible. You satisfied the requirements. That is your job. Mess it up, and correct it later, you become a hero" [17].

and use the energy of the process performers for the benefit of the client and in order to gain competitive advantage [9, pp. 77, 78].

Because of the 3rd principle, dynamic business process management is often mistaken for the real-time supervision of processes. In both cases, processes are measured and controlled during performance itself (which means e.g. ongoing control over the parameters of alert thresholds). The difference is that in the case of dynamic business process management the process performer is able to shape his or her work in a creative fashion. The monitoring of process performance does not only consist of parameter control (which is equivalent to the verification of current knowledge), but also of the identification off different changes introduce by the process performers (which is equivalent to the identification of new knowledge). First and foremost, because processes are considered completed only after having been documented, this allows us to eliminate the problem of time constraints in regard to sharing knowledge. According to the KPMG research report, this is the most severe problem associate with the issue of sharing knowledge: it is identified in as many as 64% of Polish organizations [19, p. 9]. By combining work with the uncovering of related knowledge, the 3rd principle of dynamic business process management results in the integration of day-to-day work with the process of turning tacit knowledge into explicit knowledge. In effect, the management has the possibility of broadening the concept of the learning organization to encompass all levels of hierarchy and all possible positions with an active role in performing business processes.

As an extension of traditional business process management, dynamic BPM allows for process management to encompass processes which are unstructured or even completely unforeseeable in nature [20, 21]. Dynamic business process management stipulates that employees should be authorized to introduce changes to standard, model processes in the course of performance itself (the 1st principle of dynamic BPM). Such an approach is much more adequate to the specific nature of the functioning of modern, knowledge-centered organizations, in which employees are the co-creators of knowledge and function within a double, instead of a single, learning loop [22]. In consequence, employees are empowered to use their knowledge in order to tailor a given process to the specific context of performance to the full extent of their executive privileges. What is more, by evaluating process performance with the use of e.g. techniques pertaining to big data or process mining, the organization's management is able to familiarize itself with this knowledge and the results of its use. The uncovered knowledge is documented, codified, and available to be communicated throughout the organization, as well as generalized in the form of a synthesis (3rd principle of dynamic BPM) [23].

Within the concept of dynamic BPM it is, of course, possible to operate within the limits of traditional, static business process management, treated as a specific case, in which there are no deviations from the standard process "as of today" [24, pp. 50, 51].[4] It is also possible to operate in accordance with the principles of even

---

[4]"A similar process can be observed in the case of other scientific revolutions: the past is not dismissed as "regressive and false," but rather, it becomes a part of the new paradigm as a special case" [24, pp. 50, 51].

the most rigorously understood case management. When the course of the process is completely unforeseeable, and attempts at modeling process diagrams will not provide any added value whatsoever, one should simply adopt a different form of process description[5]—however, in accordance with the principles of dynamic BPM, one should perform the process in a manner which is adequate to the situation at hand, while drawing on the experience of the knowledge worker. Of course, apart from the aforementioned extreme cases, dynamic BPM allows for actions in all intermediate cases. Using tools which allow for the quick uncovering and use of knowledge without the necessity of preparing process diagrams might turn out to be much more effective than current process tools used in e.g. crisis management processes or therapeutic processes.

2.3 The Consequences of Dynamic Business Process Management

The concept of dynamic business process management is an extension of traditional business process management. It retains all of the standard possibilities offered by traditional BPM, as well as opens process management to new, hitherto unavailable possibilities. At the same time, it allows for the creation and verification of new solutions, which open up new possibilities of practical operation, as well as new paths of development for management methodologies and their supporting IT tools.

2.3.1 The Empowerment of Process Performers

The actual source of all further possibilities brought about by adopting dynamic business process management is the reinstatement of the principle of responsibility for the work performed. The actual delegation of privileges in accordance with the 2nd principle of dynamic BPM, combined with the use of the accumulated information and its honest, objective, and quick evaluation in accordance with the 3rd principle of dynamic BPM, have the goal of creating mechanisms for motivation and evaluation, allowing for the use of the dynamism of the widest possible range of process performers.

In the knowledge economy, the most important asset of the organization are no longer fixed assets or even financial capital, but rather, the actual means of production, that is, the knowledge workers themselves and their intellectual capital. By empowering the employees to creatively shape their work in accordance with the

---

[5]In such a situation e.g. BPMN 2.0 allows for the creation of a process diagram which is limited to a single subprocess, one which includes the recommended actions ad hoc. In such a case, however, it is easier to forego modeling process diagrams and prepare a process description in the form of e.g. a checklist supporting its performance.

2nd principle of dynamic BPM ("process execution should guarantee evolutionary flexibility"), the management of the organization is operating in a consistent and commonsensical fashion [25, pp. 223, 224]. It is in the well-understood best interest of the organization for employees to be process *performers* instead of just doers; individuals who passively reenact mechanical tasks. After all, the latter can be left to industrial robots [11, pp. 38–43]. The fundamental factor which hinders innovation in traditional business process management lies in ridding process performers of their responsibility for tailoring the process to the specific context of performance. After all, it is easier and safer for the employee to perform the process in accordance with the standard. The burden of responsibility for eventual failure is placed on the creators of the standard process ("What a great catastrophe—but at least it went according to procedure!"). The 2nd principle of dynamic BPM stipulates that process performers are no longer able to rid themselves of responsibility for the effects of the performed processes—providing them with a basic impulse to conduct experiments and create or search for new knowledge. Of course, the role of the management is to help create and maintain an organizational culture and to offer additional motivators, so that the focus on innovation is even more intense. Thanks to the 3rd principle of dynamic BPM ("processes are considered completed only after having been documented"), the management has ongoing control over the introduced changes. It not run the risk of throwing the organization into chaos, because in relevant circumstances it is possible for process owners and people responsible for specific process performances to intervene at first sight of trouble.

2.3.2   The Radical Acceleration of Adapting to Requirements

The concept of dynamic BPM shifts the management of and the responsibility for performing work and its results on the direct performers, who possess the relevant knowledge on the methods of performance. It is not a coincidence that in a book devoted to speed as the decisive advantage of the enterprise do we find an explanation of why decisions should be made as close to the place of actual action as possible, if they are to be quick and to the point [26, p. 119].

One consequence of accepting the 2nd principle of dynamic BPM is shifting the decision on the specific methods of performance to the actual area of performance itself. In this concept, all of the employees is responsible for and capable of making individual attempts at innovating the process within the limits of one's executive privileges. The employees no longer has to wait for and cannot shift the responsibility on the upcoming ISO audit, the annual process overview, or a meeting of a process committee, during which all of the proposals of changes will be evaluated and decisions will be made in regard to the "correct" approach (as far as decisions can be "correct" to begin with without direct contact with the client, after the fact, and without direct responsibility for the results, etc.) This allows the employees to introduce their adaptations during performance itself (that is, during process execution), without running into further delays. At the same time, implementing the 3rd principle of dynamic BPM results in the uncovering of the effects of using current knowledge and the creation of new knowledge within the organization. This prevents the emergence of the so-called hidden factory problem known from SixSigma [27, p. 249; 28]. The problem consists of employees developing tacit processes and systems of operation, which in the best circumstance serve to alleviate the mistakes and correct the irrational nature of the "official" standard processes and procedures. Such actions are hidden from the management and hinder the possibility of introducing innovations, as in effect the management has access to inaccurate knowledge on the course, costs, and efficiency of processes.

In dynamic BPM, knowledge (new knowledge or verified older knowledge) is revealed and documented during process performance itself. In effect, it is readily available to be used throughout the entire organization. The pace of dissemination is not dependent on the decisions of the management, because thanks to IT systems (such as process portals, process mining, or elements of artificial intelligence present in BPMS and CMS systems) all employees have (or could have) practical ongoing access to documentation on the specific performances of processes and their performance indicators. There is no place for indecisiveness in regard to whether or not to accept the observable changes ("This cannot be true" vs. "We must face the facts").

2.3.3 A System of Constant Knowledge Creation and Verification

In order to develop in a time of uncertainty and turbulence, companies must achieve the same efficiency in their regeneration of processes as they have in terms of creating products and services [29, p. 68]. However, this efficiency cannot only be the result of great proficiency in predicting the foreseeable future. Organizations must combine the periodical and mostly top-down initiation of changes with a systemic mechanism of constant process improvement. By conducting limited experiments grounded in the 2nd principle of dynamic BPM, process performers are able to introduce obvious process innovations. This improvement does not take the form of monthly or quarterly action, but rather, it is an ongoing process which accompanies day-to-day operations. This internal mechanism allows for the collection and constant evaluation of the organization's market experiences, in order to enable the earliest possible identification of the undergoing changes, even those which were unforeseen in the planning stages. By evaluating the introduced modifications, their results, and the ensuing trends, the organization is able to create an early warning mechanism, a sort of "peripheral vision"[6] with a focus on upcoming changes, looking beyond whatever the management is focused on at present. It should be stressed that knowledge is created and verified in the course of holding fundamental operations, and not within a detached, separate system. In effect, there is no threat of the knowledge being or

---

[6]Peripheral vision (side vision)—all that is visible to the eye outside of the central area of focus. (From http://www.dictionary.com/browse/peripheral-vision [10.07.2017]).

becoming inconsistent or incomplete due to it being detached from the actual operations of the enterprise. This ability to constantly create and verify knowledge, along with the possibility to freely search for solutions (by means of limited experimentation), is the fundamental skill which allows the enterprise to retain its ability to both change and react in the face of changes. It also allows for the verification of the planned performance of processes in the course of day-to-day, actual operations. This day-to-day verification is fundamental. Without it, in the time of rapid and often unforeseeable changes in terms of client needs and expectations, it could turn out that the organization is falling back on old and outdated knowledge, which should have been discarded: or entered in the catalog of wrong or forbidden practices.

2.3.4   The Systemic Dissemination of Verified Knowledge

Following the implementation of dynamic business process management, there is no separation between work and sharing knowledge, as work is only considered completed after having been documented, that is, when knowledge arising from the work at hand is codified and disseminated throughout the organization (the 3rd principle of dynamic BPM). This systemic solution, which follows from the concepts of Senge [22, p. 164], enables us to overcome the fundamental practical problem tied to sharing knowledge—the lack of time or the lack of willingness to uncover knowledge by entering additional data into an additional system. However, because documentation is an integral part of ongoing, day-to-day work, without sharing knowledge it becomes impossible to perform the assigned work and reach the expected goals.

2.3.5   The Learning Organization

In the concept of dynamic BPM, knowledge management is an integral part of day-to-day operations. Because work is considered completed only after having been documented, virtually the entire knowledge of the organization becomes readily available. Enterprises managed in accordance with dynamic business process management fulfill the conditions of a learning organization from the very start. All of the employees participate in the creation of collective, explicit knowledge by identifying new solutions. One fundamental part of this process is the daily verification of current knowledge (the 2nd principle of dynamic BPM). Without it, in a time of rapid technological changes and changes in the competitive environment, by cooperating with multiple clients and taking part in multiple investments, the enterprise could face problems resulting from the use of old and outdated knowledge. The ability to constantly verify and create knowledge associated with business operations is a fundamental skill, which allows companies to retain the ability to both change and react to changes [22, pp. 19, 20].

This allows for the ongoing collection of knowledge in the form of a database of good practices and a database of wring practices, to learn not only from successes, but also from failures. As has been noted in the beginning of the second part of this chapter, dynamic business process management retains all of the possibilities of traditional business process management, while also enabling us to build and develop the institutional ability to perceive signals coming from within the organization and from the market itself, as well as the ability to adapt, so as to reach to changing client needs and expectations as quickly and as efficiently as possible. In effect, we may speak not only of the continuity of the business process, but also of the continuity of creating, obtaining, and verifying knowledge.

2.3.6 Changes to the BPM Lifecycle in the Organization

The concept of dynamic business process management requires the reinterpretation and the extension of the standard business process lifecycle (in short: BPM Lifecycle) within the organization.[7] Figure 2.1 shows an example of a process lifecycle.

Fig. 2.1 The process lifecycle in the organization. *Source* Dumas et al. [30, p. 21]

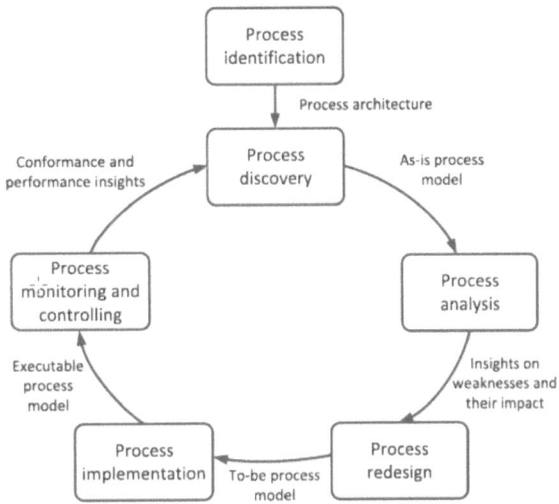

---

[7]At first, literature used the term *process lifecycle* to describe the identification, implementation, and ongoing improvement of a single process, analogous to the Deming Cycle. However, due to the identified necessity to approach the process lifecycle from the perspective of implementing and performing multiple processes in the organization, currently the term *process lifecycle* is used to encompass beside the lifecycle of a single process also actions which from the perspective of the organization prepare the implementation of process management. This "global" lifecycle in the organization will be called the business process management lifecycle (in short BPM Lifecycle).

Other similar diagrams of process lifecycles prepared on the basis of traditional business process management by software vendors, consulting companies, implementation companies, and academic researchers include analogous stages, such as:

- strategize, design, implement, compose, execute, monitor & control [31];
- model, implement, execute, monitor, optimize [32];
- define, model, simulate, implement, execute, monitor, analyze, optimize—Gartner [33];
- define, model, execute, monitor, optimize [34];
- discovery and model/remodel, validation and simulation, deployment and execution, monitoring and performance management, improve [35, p. 2];
- model, simulate, implement, deploy, execute, monitor, optimize [36, p. 499];
- (re)Design, configuration, enactment, diagnosis [37, p. 271];
- analysis, design and modeling, implementation, monitoring and controlling, refining and planning [38].

The diagrams were formed on the basis of E. Deming's PDSA cycle, which had been created over 50 years ago. It is often supplemented with additional "modern technical elements," such as simulation, implementation, reporting, etc.

To generalize: within traditional business process management, the process lifecycle can be described as a sequence consisting of sequentially executed stages with the aim of:

1. Defining goals and planning the project

In this stage, the aim is to define the aims and the methods of managing processes in accordance with the organization's strategy and its process maturity, as well as to prepare a corresponding project plan for process management within the organization.

In traditional models that are commonly found in literature, this stage often contains or is defined as: *process identification, strategizing, vision, initial process planning and strategy*, etc.

2. Designing processes

In this stage, descriptions are prepared for existing processes (*as is*) in the organization, which are then analyzed on the basis of the data in the organization, and, first and foremost, the knowledge of the personnel. In result of this analysis, an improved process model (*to be*) is prepared.

In traditional models that are commonly found in literature, this stage often contains or is defined as: *identification, discovery, defining, modeling, formalizing, simulation research, process optimization*, etc.

3. Process implementation

In this stage, the organization's operations are accommodate to the designed process model. This accommodation encompasses both training sessions and changes to the work method of the personnel, as well as changes to the operations of the ICT infrastructure and the IT systems of the organization, including the automation of process performance.

In traditional models that are commonly found in literature, this stage often contains or is defined as: *implementation, composition, positioning, process automation,* etc.

4. Process performance and monitoring

In this stage, business operations are performed and monitored in accordance with prepared and implemented process descriptions. It is becoming increasingly more common in this stage to use technologies and analytical tools from the fields of *Big Data*, process mining, robotic process automation (RPA), machine learning (ML), artificial intelligence (AI), or expert systems.

In traditional models that are commonly found in literature, this stage often contains or is defined as: *performance, monitoring, control, measurement,* etc.

5. Process improvement

In this stage, process performance is evaluated and process descriptions are improved with the aim of raising efficiency, minimizing risks, etc. At this point, techniques and analytical tools are used from the fields of *Big Data*, process mining, artificial intelligence, and expert systems.

In traditional models that are commonly found in literature, this stage often contains or is defined as: *analysis, diagnosis, optimization, improvement,* etc.

As a matter of course, the goal-defining and project-planning stage is executed once at the onset of the project of implementing business process management. The subsequent stages are performed cyclically in an unending, periodical cycle of process management and improvement. The pace of performance is dependent on the decisions of the management. It might be the result of:

- broader management plans (e.g. annual reviews);
- certification schedules imposed by external organizations (e.g. audit and certification schedules in accordance with field standards or ISO);
- unusual events (e.g. updating or preparing a new strategy or corrective actions resulting from the identification of anomalies, bad results, the development or acquisition of new technologies).

Among the different process lifecycle models in traditional business process management of note is the process lifecycle proposed by the Harvard Consulting Group. It stipulates a separate process lifecycle stage—process discovery (which consists of traditional process identification supplemented by the possibilities offered by process mining). It also defines two separate stages of process improvement named Improvement and Continuous Process Improvement [39]. However, this supplementation does not change the essence of depicting a business process lifecycle (BPM Lifecycle) within traditional process management as a sequence of stages executed one after another, preceded by a one-off execution of the preparatory stages, which initiate the implementation of process management in the organization. If we changed the symbols of subsequent stages on a diagram depicting the process lifecycle and

substituted them with the subprocess symbol known form the BPMN notation, the
process lifecycle (presented in Fig. 2.1) would depict a usual sequential "relay" process with a single feedback loop, the aim of which is to provide periodic analysis
and improve the process models on the basis of information on process performance
(Figs. 2.2 and 2.3).

In accordance with the principles of traditional process management, a process
cannot be improve or changed at all during performance itself. This can only be done
upon completion and following analysis. Processes are executed in accordance with a
description, or rather, an "algorithm," which has been prepared prior to performance.
In traditional business process management, one consequence of this principle is no
possibility of the rapid use of knowledge acquired by process performers in the course
of performance itself. This means that traditional process management may make
use of the concept of robotic process automation (RPA), but cannot operationally use

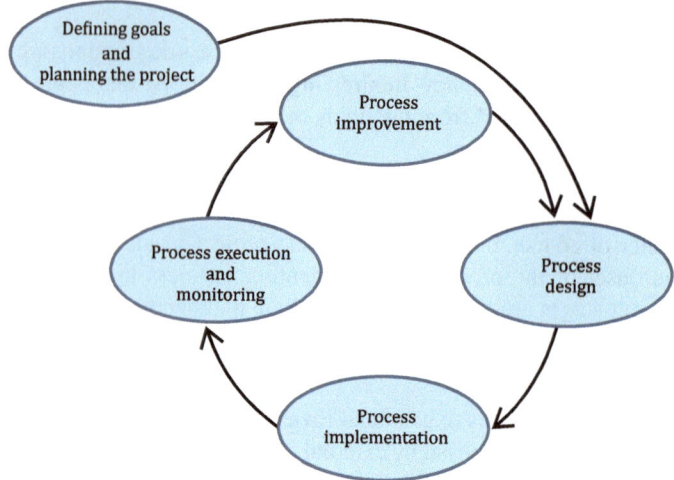

Fig. 2.2 The standard process lifecycle in accordance with traditional business process management. *Source* Author's own elaboration

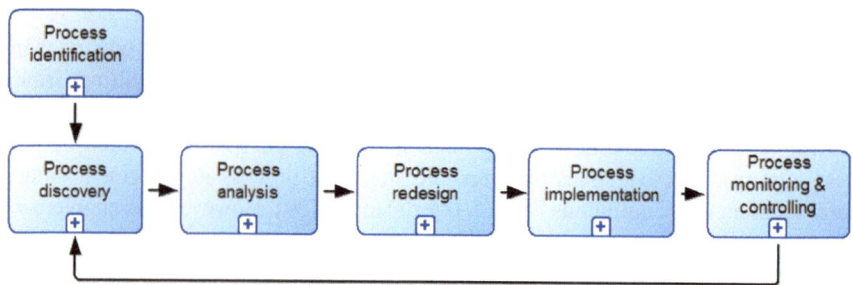

Fig. 2.3 The process lifecycle as a diagram in the BPMN notation. *Source* Author's own elaboration, on the basis of: Dumas et al. [30, p. 21]

new technologies such as machine learning or artificial intelligence in the course of process performance.

A truly meaningful qualitative change to the BPM Lifecycle was proposed in 2012 by the authors of the Process Mining Manifesto participating in the IEEE Task Force on Process Mining (Fig. 2.4) [40, p. 5].

This model has been supplemented with an *adjustment* loop, the aim of which is to adapt the process in the course of performance itself. However, the adjustment loop does not just bring about beneficial results. Sometimes, attempts at adapting the process (experiments)might result in failure or even be detrimental to the process itself. For this reason, the loop should be called "experiments with the aim of improvement." However, regardless of the name itself, the authors of the Process Mining Manifesto have stressed that organizations should include the possibility of changing ("improving") processes during performance itself as early as in the stage of designing processes and their supporting IT tools. A (re)configuration stage has also been added, in which changes are made to process-based executive systems (e.g. BPMS, document management, or workflow management), without having to repeat the implementation stage performed e.g. as the result of creating separate process performance scenarios. It has been clearly underlined that in the (re)Design stage, analysis is held in the form of e.g. simulation research on the proposed process model or in the form of comparative analyses of the new process pattern with data on completed performances (researching compliance or extending the model as the result of process mining research), with the end result being the redesigned and reconfigured process.

This is a clear step toward changes to the business process lifecycle, which allows for the dynamic management of processes. Having the option to improve on processes in the course of their performance in the form of fixes, updates, adaptations, or

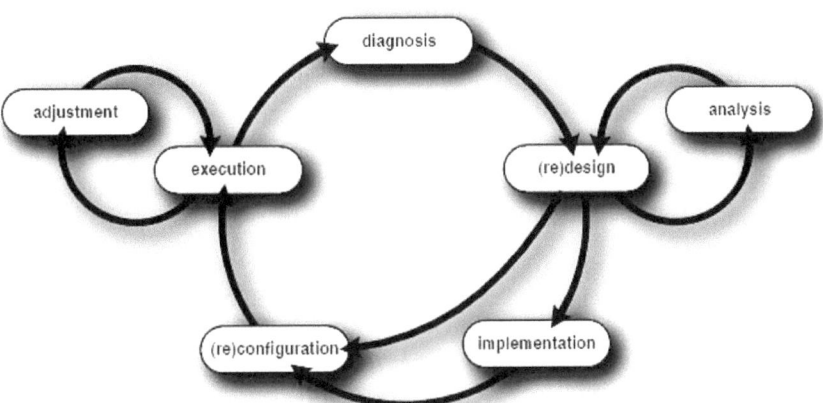

Fig. 2.4 The process lifecycle in accordance with the process mining methodology. *Source* Process Mining Manifesto [40, p. 5]

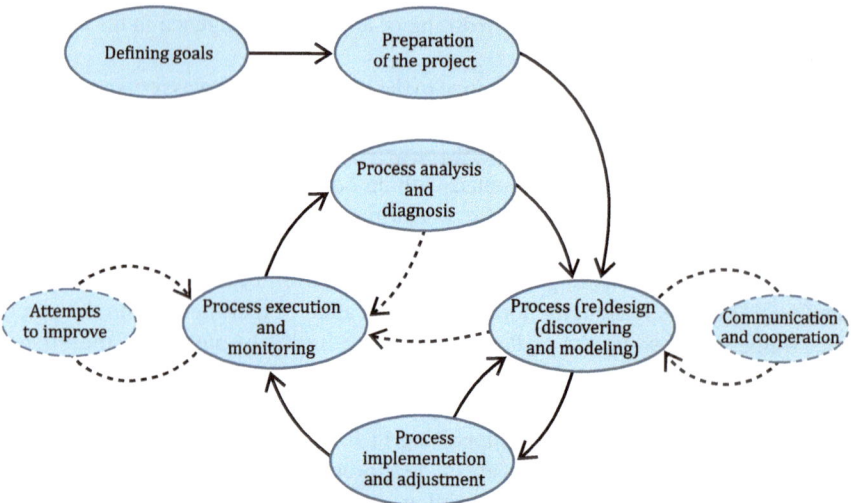

Fig. 2.5 The process lifecycle in accordance with dynamic business process management. *Source* Author's own elaboration

limited experiments, provides the process performers with the power of verifying and creating new knowledge in the course of their work. At the same time, the analysis of process performances in the (re)Design stage allows for the uncovering of such knowledge thanks to process mining. For full compliance with the concept of dynamic business process management, it is essential to manage the uncovered knowledge though the systemic combination of revealing knowledge with its evaluation and dissemination. This, however, requires us to take the concept of process lifecycles in a direction in which the performance of a process will not be equal with the perfect repetition of the standard, but rather, the repetition or adaptation of the standard with the best possible results in mind, in a manner that is the most adequate in a given context and within the limits of the executive privileges of the performer. This adaptation may be introduced by:

- process performers;
- process performers with the use of ICT solutions (e.g. online machine learning);
- autonomic elements of artificial intelligence (AI).

The postulated changes have been introduced in the process lifecycle model designed by the author in accordance with the dynamic BPM concept. The model is presented on Fig. 2.5.

The subsequent stages of the lifecycle of dynamically managed business processes are as follows:

1. Defining goals

In this stage, the goals of the project of implementing business process management are defined and agreed upon with the involved interested parties. First and foremost, they must be consistent with the organizational strategy and the actual requirements of the market. In this stage, what is defined, agreed upon, verified, and aligned with strategic goals are the goals of megaprocesses and the goals of knowledge management within the organization, as well as the rules of the implementation process, including the required level of engagement of the organization's management.

The result of this stage is: **the definition of goals and a map of the processes** in the organization (and the de facto decision on initiating the implementation of process management).

2.   Project preparation

In this stage, the organization is prepared to implement process management by:

- training sessions for the management and the employees of the organization;
- defining or verifying the organization's process maturity [41, pp. 85–87];
- verifying or preparing a detailed description of the ICT architecture in the organization;
- choosing a method of process description and communication which is the most suited to the character of the performed processes;
- preparing and implementation plan which takes into account the organization's process maturity, technological maturity, and culture;
- preparing and authorizing a document initiating the project.

The execution of each of the aforementioned tasks may result in the need for further actions to prepare the organization to implement process management in the cultural, organizational, and technological spheres.

The result of this stage is: **process architecture and a plan of implementation** accounting for the process maturity, the technological maturity, and the culture of the organization.

3.   (Re)design

In this stage, process descriptions and their data are created or updated in accordance with the rules of the organization. Process discovery is performed with the use of:

- the knowledge of the employees;
- process mining (discovery);
- the standard models for the field in which the organization operates.

Depending on the level of dynamism of the processes involved, process descriptions may take the form of:

- for static processes—detailed process models (diagrams);
- for dynamic processes—collections of tasks to be accomplished during process performance (e.g. in the form of an ontology), as well as the data required during the decision-making process and in the documentation stage.

As has been noted before, because the implementation of dynamic business process management is impossible without sufficient IT support, this stage should also include the preparation of process-centric application prototypes, which in the least should include the design and the information content of the user interface, the possible range of standard reports, and the scope of integration with ICT infrastructure. This stage should also include the preparation or implementation of new updates to existing process-centric applications supporting the performance of business processes in the organization.

Furthermore, in this stage, the organization's internal rules and regulations should be—where required—updated for consistency between dynamic business process management with other fields of management.

– Communication and cooperation

In the (re)Design stage—in accordance with the principles of dynamic business process management—in order to make good use of the broadest possible part of the organization's intellectual capital, the proposed process descriptions, prototypes, or applications which have been cleared for testing should be consulted with in-house and outside experts, and, first and foremost, with practitioners themselves, who use them on a daily basis, through e.g. communities of practice or social media websites.

4. Implementation and adjustment

In this stage, process descriptions are implemented (and eventual changes to other internal regulations are introduced) along with their supporting process-centric applications within the organization, with the engagement of and a focus on:

- the organization's management: imparting knowledge on topics pertaining to process goals, the main fields of changes to past methods of operation, methods of monitoring and the expected efficiency of process performance, as well as the main risks and methods of reacting in abnormal and risk-prone situations;
- the knowledge workers directly involved in process performance: imparting knowledge on topics pertaining to process goals, the main fields of changes to past methods of operations, the expected efficiency and indicators of process performance, the scope of authorizations to dynamically shape the performed processes, the sources of knowledge available in the organization and processes of knowledge dissemination, methods of operation in situations which demand adapting (changing) the course of a process, as well as the main risks and methods of reacting in abnormal and risk-prone situations.

In this stage, it is possible to adapt process descriptions and the configurations of their supporting IT systems to needs and requirements required during implementation. Should it turn out that a designed process or a configuration of a process-centric application does not meet the expectations of the users, it is possible to return to the (re)Design stage in order to prepare the process descriptions and applications once again.

5. Execution and monitoring

In this stage, business processes are performed, and data on their performance is collected on an ongoing basis. For transaction systems (e.g. ERP or EMR) and workflow systems (BPMS) they are stored in event logs. Data from other sources (e.g. mobile applications, social media applications, e-mail accounts) should be integrated within a unified data source. Such information should be monitored by control systems on an ongoing basis, as well as analyzed and used in the ongoing support of knowledge workers by software elements of artificial intelligence [6, p. 38].

– Attempts at making improvements

In accordance with the 2nd principle of dynamic business process management, knowledge workers have the power to create or adapt described business processes to the requirements of a specific context of performance. In accordance with the 3rd principle of dynamic BPM, the undertaken attempts at improving the processes should be documented and monitored on an ongoing basis.

6. Analysis and diagnosis

In addition to business processes being monitored in the performance stage, they are nevertheless evaluated ex post by means of:

- standard control actions, including the control of process efficiency, duration, costs, resources used, risks involved, etc.;
- uncovering the actual course of the performed processes and evaluating the results of the implemented improvements with the aim of:
  - broadening the processes of the organization through communication (adding to the list of best practices and informing about the update), as well as redesigning and tailoring processes and their supporting applications;
  - communicating information on the negative results of a specific attempt at improving a process (adding to the list of wrong practices and informing about the update);
  - initiating a broader evaluation of the possibilities of using a discovered potential improvement (while informing the relevant parties about the option to participate in the discussion).

Knowledge obtained in this stage should be systematically communicated to authorized members of the organization, with a particular focus on the employees who are directly responsible for process performance, for whom new or verified knowledge might have direct significance (in the performance and monitoring stage). This requires the existence within the organization of a culture and mechanisms of internal communication, which allow for the ongoing, broad improvement of processes and dissemination of knowledge, as well as the existence of an ICT infrastructure enabling the rapid introduction of changes and rapid communication.

At the same time, improvements resulting from practical attempts at innovation, which have been given a positive evaluation, within the developed process lifecycle resulting from dynamic business process management may be introduced in the performance and monitoring stage directly following the (re)Design stage, without the necessity of going through the implementation and adjustment stage. As previously, this requires organizations to develop efficient mechanisms of internal communications both on the level of social culture and work culture, as well as on the level of ICT infrastructure, understood as e.g. the broad acceptance and the efficient use of mobile devices, social media applications, or elements of artificial intelligence. It should be noted that in the proposed process lifecycle, improvements may be introduced in the course of the following three stages:

1. The (re)Design stage

Processes are improved on the basis of knowledge uncovered in the analysis and diagnosis stage or created or obtained by way of communication and cooperation. The improvement process can be performed:

- on the initiative of the management of the organization in the form of one-off actions (e.g. resulting from changes in ownership or changes to the organizational strategy) or periodical evaluations (audits) performed in accordance with implemented quality assurance methodology and internal and external rules and regulations;
- on the initiative of process performers and managers, resulting from the ongoing evaluation of results and process performance pathways. Improvements are being introduced within the limits of executive privileges.

The process should engage all of the knowledge workers participating in the performance of a given process or whose processes are in direct communication with the process in question.

2. The implementation and adjustment stage

Furthermore, in the implementation and adjustment stage both process descriptions, as well as their supporting process-centric applications can be tailored to the needs and requirements identified in the course of implementation. This adjustment may either require a broader implementation or merely proper communication within the organization.

3. The execution and monitoring stage

Processes are adapted or created to fulfill the requirements of a specific context of performance by their knowledge workers. At the same time, the knowledge uncovered or created in the course of performance is evaluated by the client and communicated throughout the organization. In the case of dynamic business process management, this process is ongoing irrespective of the plans or the expedient actions of the organization's management.

The proposed process lifecycle is consistent with case management methodologies, or rather, it is their extension. The (re)Design stage extends the possibilities of the case management modeling stage. The performance and monitoring stage corresponds with the discovery stage, while the analysis and diagnosis stage along with part of the (re)Design stage corresponds with the adaptation stage. The BPMS systems available to date allow for the use of methodologies and forms of process description used in both case management and dynamic business process management in different parts of a single process-centric application in accordance with the proposed model. In effect, the described dynamic BPM process lifecycle can be implemented in practice with the concurrent use of elements of process management/case management methodologies, depending on the specific nature of the tasks and subprocesses involved.

2.3.7   The Necessity of Supplementing Business Process Maturity Models

Dynamic business process management requires us to supplement existing models of process maturity within the organization. The most known models of process maturity define from five (e.g. CMM, BPMM, or BPOMM) to seven (Nowosielski) levels of maturity [18, 41, pp. 75–95; 42, 43; 44, pp. 144–146] (Table 2.1).

In the case of all of the included models process maturity levels 1, 2, 3 correspond to the shift of management in the organization from a state in which processes are unrecognized and unidentified to a state in which processes are identified, standardized, and documented. The subsequent levels 4 and 5, and in Nowosielski's model also 6 and 7, correspond to a shift within the organization toward planning, constant improvement, and management of identified processes. Regardless of the terminological differences, in each of the process maturity models on the highest levels processes are "managed" and/or "continuously improved". On these levels:

- "all employees understand processes" (Rummler and Brache: Process Performance Index—PPI);
- "processes are ongoing and innovated upon" (OMG: Business Process Maturity Model—BPMM);
- "processes are proactively monitored and controlled" (Lee: Business Process Maturity Model—BPMM—Lee);
- "the organization has introduced productivity management along with a cause analysis and ongoing improvement" [Capability Maturity Model Integrated (CMMI)];
- "the organization cooperates with the suppliers" [BPO Maturity Model (BPOMM)—McCormack];
- "the first iteration consists of identifying processes within the enterprise, evaluating process and making management decisions on the basis of performed analyses (these decisions mostly pertain to improving the analyzed processes). Each subsequent iteration consists of the measurement of processes, the analysis of the data obtained therefrom, and potential further improvements" (Procesowcy).

What follows from the aforementioned models is that organizations which merely use a fraction of their intellectual capital (just the management, process owners, or process leaders) could nonetheless still be judged to be on the highest levels of process maturity, while the constant improvement of processes is performed within a traditional process with top-down management. In other words, an organization which is deemed to possess the highest process maturity level may nonetheless find itself unable to cope with the individualization of client needs, nor with the constant uncovering of tacit knowledge. And since in the knowledge economy those

Table 2.1 Comparison between selected models of process maturity within the organization

	Software Engineering Institute (SEI)	Object Management Group (OMG)	McCormack	Dumas, La Rosa, Mendling, and Reijers	Nowosielski
	Capability Maturity Model (CMM)	Business Process Maturity Model (BPMM)	BPO Maturity Model (BPOMM)	Capability Maturity Model Integrated (CMMI)	
7					Process management
6					Systematic process improvement
5	Constantly improved processes	Constantly improved processes	Extended processes	Processes optimized on an ongoing basis	Process planning and control
4	Processes managed on the basis of metrics	Predictable processes	Integrated processes	Processes managed on a qualitative basis	Measurement and registry of process results
3	Organized and identified, but unmeasured processes	Standardized processes	Interconnected processes	Defined processes (regardless of organizational silos)	Precess identification and documentation
2	Repeatable processes, partly organized	Repeatable processes	Defined processes	Projects and processes planned and managed in practice	Creation of the awareness of the need for a process-based approach
1	Unrecognized and unorganized processes	Unrecognized processes, *ad hoc* actions	*Ad hoc* processes	Initial, processes performed on an *ad hoc* basis	No awareness of the need for a process-based approach

Source Author's own elaboration on the basis of i.a. McCormack and Lockhamy [45, p. 5], OMG [18], Röglinger et al. [42], Bitkowska [44, pp. 144–146], Dumas et al. [30, pp. 39, 40]

and other factors, which are nonexistent in traditional process management, may be decisive in terms of the potential success of the organization—or even its continued existence—dynamic business process management stipulates the need to supplement traditional models of process maturity with additional parameters (e.g. the use of intellectual capital or the efficiency of knowledge management). Stefan Nowosielski diagnosed this need the most effectively. When speaking of the highest, seventh level of *Process management,* his authorial model of process maturity, he has made the following statement: "one very important component of the organization in this stage is an appropriate organizational culture along with knowledge management" [44, p. 146].

Two solutions are possible:

- models of process maturity within the organization should include an additional level (or multiple levels), on which business processes would be adapted or individualized to the specific context of performance during performance itself by their direct performers, who are provided access to the codified knowledge of the enterprise;
- the highest level or several of the highest levels of the organization's process maturity models should be reinterpreted, so that the organization's maturity level could be determined in more detail depending on the implemented mechanisms of improvement.

Regardless of the chosen extension, the organization's process maturity models will have to include the maturity level of knowledge management, and—indirectly—the maturity level of the knowledge workers themselves.

2.3.8 Change of the Goals of Implementing Process Management in Organizations

In the case of the traditional, static approach to implementing process management, the organization's managers were convinced that it is sufficient enough to design processes and then purchase and implement a system to automate their performance [46]. The aim of implementation was to simply "raise the efficiency" of the performed processes. This was usually a one-off undertaking, after which the efficiency of the processes soon returned to its usual levels, and the organization itself returned to its day-to-day routine consisting of the performance of standard processes. The abovementioned method of implementation led to the use of knowledge available at the moment of implementation or since the last update. However, the value of static knowledge falls the quicker, the quicker the changes in the organization's environment. Under the conditions of hypercompetition, the value of knowledge might even be negative, as old, outdated knowledge might hinder the creation of new products and services [14]. In the most optimistic cases, in the scope of traditional implementations of process management organizations accounted for the possibility of the continuous improvement of processes on the part of the management—in this context called process "owners" or "leaders." However, the companies were still in the situation of using just a mere fraction of their entire intellectual capital, as new knowledge creation in the organization was filtered by the management, whose knowledge alone formed the basis of current business processes.

As has been stressed above, managers are becoming increasingly more aware of the drawbacks of such an approach to process management. Increasingly more often the goal is no longer for the process performers to enact optimal ("ideal") processes defined by the company's management. Instead, the goal is the adaptation, or the dynamic adjustment, of processes to changing client needs and changing rules of competition. This, in turn, requires having access to current and practicable knowledge. Ideal business processes no longer exist, but a new approach to implementing process management is still essential [47, 48, p. 26].

2.4 New Possibilities of Using ICT Technologies and Methodologies in Dynamic Business Process Management

As has been demonstrated in Chap. 1, most of the organization's processes in the knowledge economy fall outside the scope of traditional business process management. The situation has changed considerably following the extension of process management to dynamic business process management, which resulted in the necessity (or perhaps provided the chance) to put ICT methodologies and technologies to better use. Their use has been impossible or ineffective thus far: why, after all, implement process mining techniques or online machine learning under the conditions of traditional business process management, when we are not in fact looking for innovative deviations from the standard process? In order to punish the employees for innovations deviating from binding procedures and rules and regulations? Why, after all, implement the collection of data resulting from informal information flows in the course of process performance (social BPM), when unforeseen social interactions are not included in standard models and evaluations, despite having a potential strong effect on business decisions?

2.4.1 ICT Technologies and Business Process Management

In the case of traditional business process management, processes were first built into ("embedded") transactional applications, such as MRP II, ERP, CRM or HIS systems. In the traditional concept, there is no need to regularly change processes or create process scenarios tailored to the needs of different groups of clients or users. Unfortunately, even in the instance when processes are changed rarely did their entry into transactional systems require the initiation and maintenance of complex, expensive, and often long-term programming projects. Another attempt consisted of supplementing existing transactional systems with modules dedicated to describing, simulating, optimizing, and communicating business processes [49, 50]. The standardization of notation (e.g. BPMN) and the language of process automation (Business Process Execution Language—BPEL or Business Process Execution Language for Web Services—BPEL4WS) has led to the availability of solutions allowing for the flexible combination of applications dedicated to process description with transactional and analytical systems [51]. At the same time, we witnessed the development from the ground up of Business Process Management Systems (BPMS), which allowed for the modeling and simulating of business processes, their performance and evaluation, as well as their integration with existing MRP II, ERP, CRM, or EMR systems. At present we are increasingly witnessing the emergence of new versions of ERP and CRM systems with built-in modelers, which are designed around defined business processes. Just like BPMS applications, they enable the modeling and simulation of business processes, as well as their performance and evaluation [52–54].

In order to enable the implementation of the 2nd and 3rd principles of dynamic business process management, it is essential for IT systems, regardless of their path of origin, to enable:

- the flexible adaptation of business processes by process performers in the course of performance itself;
- support for knowledge management, and in particular, the option to collect and evaluate incoming knowledge on an ongoing basis, including data on its full context and the individuals participating in its creation;
- the distribution of authorizations and responsibilities for administering and performing processes [55, p. 9].

The company Ultimus made such an attempt as early as in 2005. In Ultimus BPM Suite 7.0, which offered Adaptive Discovery functionality, it is possible [56]:

- to limit the identification of processes to preliminary process modeling or modeling their known, permanent elements with the use of the system's "learning" mechanism;
- for process performers to introduce changes to processes in the course of performance itself.

Ultimus Business Suite 7.0 was a system, which from a technological standpoint seemed to adhere to the principles of dynamic business process management. However, in practice it turned out that knowledge workers did not accept the proposed method of creating or modifying process diagrams "on the fly." In conclusion, it is fundamental to polish the method of process description and the form of process presentation, as well as the user interface of the system. It is not surprising that according to Forrester [57], the leading suppliers of BPMS and CMS systems devote 50% of their efforts and budget to this very issue. This and other attempts at tailoring BPMS systems to the dynamic management of unstructured, ad hoc, individual, and unforeseeable processes, which have been made in the last decade, have led to the closer integration of BPMS and CMS systems, as described in Sect. 1.4.6.

2.4.2   The Practical Reunification of Process Management with Case Management—An Attempt at Surpassing the Limitations of Traditional Business Process Management

There are no areas of business, in which there would exist only "process-centric" or "case-centric" companies. No such classification of organizations into two categories exists at all. However, in consequence of the limitations of traditional business process management described in Chap. 1, such a classification emerged in regard to IT systems themselves. However, from a business perspective, supporting management with the use of IT systems is simply beneficial or ineffective—IT tools either support management or hinder it. The leading business managers are not really interested in "what color is the cat?", but rather, "does the cat hunt mice?" For this reason, as has been demonstrated in Chap. 1, the expectations and pressures of business have lead to the closer integration, or even the interfusion of the methodologies and tools pertaining to business process management and case management, as has been noted by Gartner and Forrester.

As Fig. 2.6 demonstrate, in all two research projects performed by Gartner, which encompassed various defined categories of IT systems, in 2016 the leaders both in terms of current functionality, as well as in terms of the completeness of vision in regard to the created solution, were Pegasystem, Appian, and IBM. Exactly the same companies are in Forrester Analysis of the market for Dynamic Case Management Systems in Q1 2016 [57]. Other high-ranking solutions were products belonging to Column Technologies, DST Systems, Lexmark, MicroPact, Newgen Software, OpenText, and Tibco. Despite the ongoing status of theoretical discussions on the scope of using the concept of dynamic BPM in practice, no longer are there BPMS systems on offer, which would operate in accordance with traditional business process management [58–62]. Most of the systems offer functions which fall outside the scope of traditional business process management, while some combine features of both BPMS and CMS (Bizflow, Fujitsu, IBM BPM, Webmethods) [56, 63]. By analogy, most CMS systems offer the functionality of modeling and using process models in case management [64]. The integration of BPMS and CMS methodologies and tools has practically been introduced. IBM BPM (IBM) and Adaptive Case Management (ISIS Papyrus) already enable us to use the entire functionality encompassing static and dynamic processes performed in the basis of process diagrams and checklists characteristic of case management within a single process-centric application [64, 65].

This path of development has been included by the authors and is visible in Gartner's report: Magic Quadrant for Intelligent Business Process Management Suites from 2017 [61].[8] Among the critical possibilities of iBPMS systems are the following:

- Process discovery and optimization;
- Context and behavior history, which means that process exploration techniques—just like simulation research—have become an integral part of iBPMS systems.

[8]Gartner, Magic Quadrant for Intelligent Business Process Management Suite Report, published on October 24, 2017 [61].

Magic Quadrant
for Intelligent
Based Business Process
Management Suites

Magic Quadrant
for BPM-Platform
– Case Management
Frameworks

Fig. 2.6 Comparison between the Intelligent Business Process Management Suites and BPM-Platform-Based Case Management Frameworks systems in the year 2016. *Source* Gartner [66, 67]

In the years 2015–2016 the authors of the report have evaluated the following cases of using iBPM systems:

- Composition of intelligent, process-centric applications;
- Continuous process improvement;
- Process transformation;
- Process digitization.

In 2017, the report also evaluated;

- Citizen developer application composition;
- Case management.

According to Gartner, case management and empowering users with the option of introducing changes in applications are necessary components, the lack of which in 2017 disqualifies a given BPMS system from a market standpoint.

From a theoretical, but also a practical perspective, the challenge is not to establish which methodologies or tools are better and for which fields and organizations, but rather, to introduce a generalization in the form of a single, coherent concept encompassing both traditional process management and case management.

2.4.3 New Database and Internet Technologies

Due to rapid changes in social culture and the ensuing changes in work culture pertaining to the use of ICT technologies, a practical concept of dynamic business process management should also encompass and make use of new or emerging ICT methodologies and technologies, such as:

- Cloud computing;
- The Internet of Things (IoT) [68, pp. 20–25];
- Big Data [69, 70];
- Social technologies, social BPM [48, 71, pp. 66–68];
- Process mining [40];
- Robotic process automation (RPA) [72, 73];
- Machine learning and Artificial intelligence [74, pp. 230–236].

Dynamic business process management has set a clear practical goal in regard to both process mining and machine learning, as well as the research and analysis of social networks within the organization. The goal is not only to monitor deviations from the model during process performance, but first and foremost, to gather knowledge on the decisive factors behind and the results of introducing innovations to processes, as well as information on the knowledge workers who are responsible for their introduction. To this end, it is crucial to ensure the broadest possible use of processes that are strictly connected with dynamic business process management:

- social media applications and the Internet of Things with the aim of acquiring information;
- cloud computing and *Big Data* analysis with the aim of evaluating the collected information on an ongoing basis;
- robotic process automation, process mining, and machine learning with the aim of supporting knowledge workers in the course of process performance;
- artificial intelligence with the aim of autonomizing process execution.

The use of tools enabling the uncovering and quick use of knowledge without the need of preparing process diagrams might be much more efficient than existing tools pertaining to traditional business process management, in the case of e.g. crisis management processes and therapeutic processes. As the User Trained Agent (UTA) component by ISIS Papyrus shows, it is possible to leave the decision on the possible use and the scope of online help by a learning system in regard to the possibility of tailoring the performance of processes to the knowledge of the organization to the user him—or herself. The user is free to either look up the content of a given tab—or not [75].

The methodology created following the unification of methodologies and tools from the fields of process management and case management will not be fully functional without the option to manage and quickly use the knowledge uncovered and verified in the course of process performance [76–78].

2.5 Conclusions—The New Meaning of the Term "Business Process Management"

Modern business process management is dynamic in nature, because such are the requirements of business. Modern commerce already makes use of specific elements of dynamic BPM, even when still not under that name itself. The aim of the theoretical reflection presented in this chapter, which will be continued in Chaps. 3–4, is to present all of the possibilities it offers and the requirements it imposes on the organizations implementing dynamic business process management. The research presented in this chapter, as well as Fig. 2.7, clearly demonstrate that dynamic business process management has a much broader scope of application in modern organizations than its traditional, static counterpart.

Dynamic BPM is a much broader concept, which includes traditional, static business process management as a special case. Unfortunately, for years the concept of static business process management has been synonymous with the terms *process management* and *business process management*—BPM. Its as if we still used the term *automobile* to antiquated steam engine cars or still took the term *industrial production* to be synonymous with Ford's production line. Unfortunately, though this sounds archaic, even most specialists in the field of management still thinks of traditional, static process management, with all of its limitations, when using the term *process management*. In order to avoid confusion, it is essential to introduce changes to the terms involved. *Business process management* (BPM) should now be understood as *dynamic business process management* (dynamic BPM). Its special

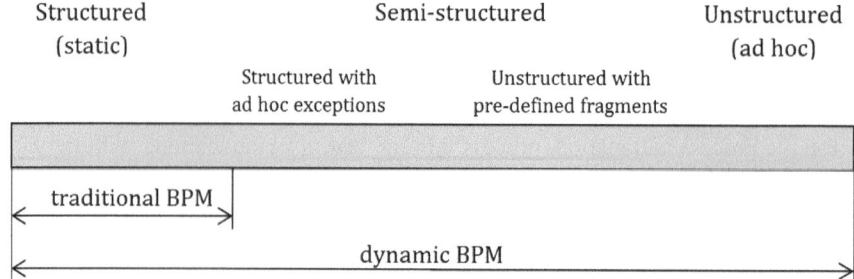

Fig. 2.7 The spectrum of processes within the organization. *Source* Author's own elaboration, on the basis of: Kemsley [79], Di Ciccio et al. [21]

case, which is still sporadically in effect, is traditional (static) process management, which thus far has been synonymous with the term *BPM*. In practice, this seemingly cosmetic change has all of the aforementioned consequences.

By answering the questions posted in the introduction to this chapter, it is possible to extend traditional business process management with a view to adapting it to the conditions of the knowledge economy. It is also possible to unify process management with case management. Developed by the author in the course of his studies, the proposal of the dynamic BPM process lifecycle is in accordance with the requirements set in Chap. 1, as well as with the practice of case management. The Defining goals, Identification, and (re)Design stages are the equivalent of the Modeling stage within *adaptive case management*, or even surpass their functional scope. The Implementation and adjustment and Performance and monitoring stages are the equivalent of the Discovery stage. The Analysis and diagnosis stage is virtually identical with the Adaptation stage. In effect, case management and (dynamic) business process management only differ in terms of the scope of practical elements of implementation within the organization:

- the method of process description and the form of process presentation,
- the ergonomics of IT systems in regard to creating and presenting process descriptions, as well as introducing changes to ongoing business processes.

In effect, it is possible to use both approaches interchangeably within the organization, depending on the nature of the processes involve, or simultaneously within a single process, depending on the type of the task or number of tasks at hand [65, 80]. In the case of static processes or static processes with dynamic exceptions, one may successfully use "process-centered" methodologies and functionalities, and in the case of dynamic processes—methodologies and functionalities originating from case management. In consequence, it is possible to unify (dynamic) business process management and case management, as there were no conflicts between the philosophies and methodologies of their implementation and functioning since their very onset [71, pp. 59, 60].[9] Instead, there was the irrational approach and the reluctance

---

[9]"I have to stress that Adaptive Processes are fully aligned with accepted management theory in terms of Business Architecture, Balanced Scorecard and BPM. I even propose that Adaptive Process is finally a process management concept that implements all functionality in technology to make BPM happen the way it was originally intended" [71, p. 58].

to face facts on the part of the "genuine process-centered experts," for whom process descriptions meant a "sequence of actions:" independent from the context of performance, possible to identify with absolute certainty, and optimized to the level of an "ideal" before performance itself.

In the case of traditional business process management, process performers were prohibited from introducing changes to the process in the course of performance itself. Their knowledge, in turn, remained unused, or wasted. **Traditional, static business process management has led to the systemic waste of intellectual capital, despite the fact that in the knowledge economy, success is impossible without the creation and day-to-day use of knowledge.** This has been envisioned as far back as in the Nineties by Peter Senge: "As the world becomes more interconnected and business becomes more complex and dynamic, work must become more "learningful," and "[t]he organizations that will truly excel in the future will be the organizations that discover how to tap people's commitment and capacity to learn at all levels in an organization" [22, p. 19]. For this reason, the most essential consequence of dynamic business process management is the inextricable connection of process management and case management.

References

1. Hammer M (1990) Reengineering work: don't automate, obliterate. Harvard Business Review, July/Aug 1990, pp 104–112
2. International Organization for Standardization (2015) ISO 9000 family—quality management. Retrieved from https://www.iso.org/iso-9001-quality-management.html [20.10.2018]
3. Tiwana A (2001) The essential guide to knowledge management; E-business and CRM applications. Prentice-Hall, Upper Saddle River
4. Wismer D (2012) Google's Larry Page: "Competition Is One Click Away and Other Quotes of the Day". Retrieved from https://www.forbes.com/sites/davidwismer/2012/10/14/googles-larry-page-competition-is-one-click-away-and-other-quotes-of-the-week/
5. Oxford Dictionary (2018) Dynamism. Retrieved from https://en.oxforddictionaries.com/definition/dynamism [11.06.2018]
6. Kemsley S (2010) Runtime collaboration and dynamic modeling in BPM: allowing the business to shape its own processes on the fly. Cutter IT J 23(2):35–39
7. Heller M, Pabjan T (2014) Elementy filozofii przyrody. Copernicus Center Press, Kraków
8. Britannica (2018) Occams razor. Retrieved from https://www.britannica.com/topic/Occams-razor [25.05.2018]
9. Rummler G, Brache A (2000) Podnoszenie efektywności organizacji (Improving performance). PWE, Warszawa
10. Nowosielski S (2012) Zarządzanie procesami. Retrieved from http://procesy.ue.wroc.pl/uploads/pliki/procesy/wyklady/ZPRnowosielskiWYKLAD.pdf [8.08.2017]
11. Hammer M (1999) Reinżynieria i jej następstwa—jak organizacje skoncentrowane na procesach zmieniają naszą pracę i nasze życie (Beyond reengineering. How the process-centered organization is changing our work and our lives). Wydawnictwo Naukowe PWN SA, Warszawa
12. Porter M (2001) Porter o konkurencji (On competition). PWE, Warszawa
13. Champy J (2002) X-engineering the corporation. Reinventing your business in the digital age. Warner Books, New York
14. D'Aveni R (1994) Hypercompetition managing the dynamics of strategic maneuvering. The Free Press, New York
15. Drucker P (2000) Zarządzanie w XXI wieku (Managements challenges for the 21st century). Muza S.A, Warszawa

16. Toffler A (1990) Power shift. Knowledge, wealth, and violence at the edge of the 21st century. Bantam Books
17. Deming WE (1986) Out of the crisis. Massachusetts Institute of Technology, Center for Advanced Engineering Study, Cambridge
18. OMG (2008) Business process maturity model (BPMM) version 1.0. Retrieved from http://www.omg.org/spec/BPMM/1.0/PDF. [29.03.2017]
19. KPMG (2004) Zarządzanie wiedzą w Polsce 2004. Raport badawczy
20. OMG (2011) Business process model and notation (BPMN). Retrieved from http://www.omg.org/spec/BPMN/2.0 [3.04.2016]
21. Di Ciccio C, Marrella A, Russo A (2012) Knowledge-intensive processes: an overview of contemporary approaches? In: 1st international workshop on knowledge-intensive business processes (KiBP 2012), Rome, Italy, 15 June. Retrieved from http://ceur-ws.org/Vol-861/KiBP2012_paper_2.pdf [2.04.2016]
22. Senge P (1990) The fifth discipline. The art and practice of the learning organization. Currency Doubleday, New York
23. Jung J, Choi I, Song M (2007) An integration architecture for knowledge management systems and business process management systems. Computers in Industry 58 (2007): 21–34.
24. Heller M (2014) Granice nauki. Copernicus Center Press, Kraków
25. Liker JK (2005) Droga Toyoty. 14 zasad zarządzania wiodacej firmy produkcyjnej świata (The Toyota way. 14 management principles from the world's greatest manufacturer). MT Biznes Ltd., Warszawa
26. Jennings J, Haughton L (2002) Szybkość jako atut w biznesie: to nie duzi zjadają małych, ale szybcy opieszałych (Its not the big that eat the small … Its the fast that eat the slow. How to use speed as a competitive tool in business). MT Biznes, Warszawa
27. Thompson JR, Koronacki J, Nieckuła J (2005) Techniki zarządzania jakością od Shewarta do metody "Six Sigma". Akademicka Oficyna Wydawnicza EXIT, Warszawa
28. BusinessDictionary (2017) Hidden factory. Retrieved from http://www.businessdictionary.com/definition/hidden-factory.html [31.03.2017]. ("Activities that reduce the quality or efficiency of a manufacturing operation or business process, but are not initially known to managers or others seeking to improve the process") [4.09.2017]
29. Hamel G, Yalikangas L (2003) W poszukiwaniu zdolności strategicznej regeneracji. Harvard Business Review, Nov 2003
30. Dumas M, La Rosa M, Mendling J, Reijers H (2016) Fundamentals of business process management. Springer, Heidelberg
31. Software AG (2011) Enterprise BPM series: a summary. Retrieved from http://www.ariscommunity.com/users/nina-uhl/2011-07-26-enterprise-bpm-series-summary [28.03.2017]
32. PNM SOFT (2017) BPM lifecycle. Retrieved from http://www.pnmsoft.com/resources/bpm-tutorial/bpm-lifecycle/ [29.03.2017]
33. Polancic G (2013) Learning BPMN 2.0—Business process vs workflow. Retrieved from http://blog.goodelearning.com/bpmn/business-process-vs-workflow/ [10.02.2017]
34. BPM Resource Center (2012) An unbiased, straightforwaed resource for business managers and analysts on BPM & BPMS. Retrieved from http://www.what-is-bpm.com/get_started/bpm_methodology.html [28.03.2017]
35. Pourshahid A, Amyot D, Peyton L, Ghanavati S, Chen P, Weiss M, Forster A (2009) Business process management with the user requirements notation. Electron Commer Res 9(4):269–316. https://doi.org/10.1007/s10660-009-9039-z
36. Gulledge T, Hiroshige S, Manning D (2011) Composite supply chain applications. Retrieved from https://www.researchgate.net/publication/221915222_Composite_Supply_Chain_Applications [8.08.2017]
37. Di Ciccio C, Marrella A, Russo A (2015) Knowledge-intensive processes characteristics, requirements and analysis of contemporary approaches. J Data Semant 4(1):29–57. Retrieved from https://www.researchgate.net/profile/Claudio_Di_Ciccio/publication/269629902_Knowledge-Intensive_Processes_Characteristics_Requirements_and_Analysis_of_Contemporary_Approaches/links/576a501a08ae1a43d23bca3c.pdf [18.07.2017]

38. Bernardo R, Ribeiro Galina S, Dallavalle de Pádua S (2017) The BPM lifecycle: how to incorporate a view external to the organization through dynamic capability. Bus Process Manag J 23(1):155–175

39. Harvard Consulting Group (2017) Business process management lifecycle. Retrieved from http://harvardcomputing.com/consulting/ [10.07.2017]

40. IEEE Task Force on Process Mining (2012) Process mining manifesto. Retrieved from http://www.win.tue.nl/ieeetfpm/doku.php?id=shared:process_mining_manifesto [02.04.2016]

41. Kania K (2013) Doskonalenie zarządzania procesami biznesowymi w organizacji z wykorzystaniem modeli dojrzałości i technologii informacyjno-komunikacyjnych. Wydawnictwo Uniwersytetu Ekonomicznego w Katowicach, Katowice

42. Röglinger M, Pöppelbuß J, Becker J (2012) Maturity models in business process management. Bus Process Manag J 18. Retrieved from http://www.fim-rc.de/Paperbibliothek/Veroeffentlicht/352/wi-352.pdf [29.03.2017]

43. Procesowcy (2013) Dojrzałość procesowa polskich organizacji. Podsumowanie II edycji badania dojrzałości procesowej polskich organizacji. Retrieved from http://carrywater.com/wp-content/uploads/2013/02/raport_dojrzalosc_procesowa_2013.pdf [29.03.2017]

44. Bitkowska A (2013) Zarządzanie procesowe we współczesnych organizacjach. Difin, Warszawa

45. McCormack K, Lockamy A (2004) The development of a supply chain management process maturity model using the concepts of business process orientation. Supply Chain Manag

46. Spanyi A (2006) The "M" is missing in BPM. Retrieved from http://www.bpminstitute.org/resources/articles/%E2%80%9Cm%E2%80%9D-missing-bpm [14.11.2016]

47. Russell Records L (2005) The fusion of process and knowledge management. Retrieved from: http://www.bptrends.com/publicationfiles/09-05%20WP%20Fusion%20Process%20KM%20-%20Records.pdf [8.04.2016]

48. Swenson K (2012) The quantum organization: how social technology will displace the Newtonian view. In: Fisher L (ed) Social BPM: work, planning and collaboration under the impact of social technology. Future Strategies Inc, Lighthouse Point, pp 19–34

49. Oracle (2009) Accelerate business processes with Oracle enterprise application documents. Retrieved from http://www.oracle.com/us/products/middleware/content-management/059454.pdf [21.07.2017]

50. Software AG (2016) ARIS for SAP solution. Retrieved from http://www2.softwareag.com/corporate/products/aris_alfabet/bpa/aris_sap/default.aspx [12.05.2017]

51. Kelly D (2004) Orchestrating business processes with BPEL. Retrieved from http://www.ebizq.net/topics/bpm/features/3966.html?page=1 [14.11.2004]

52. Microsoft (2015) Microsoft dynamics lifecycle services video tutorial. Retrieved from http://www.techbrothersit.com/search/label/Microsoft%20Dynamics%20Lifecycle%20Services%20Video%20Tutorial [21.07.2017]

53. Microsoft (2017) Business process modeler. Retrieved from https://docs.microsoft.com/en-us/dynamics365/unified-operations/dev-itpro/lifecycle-services/business-process-modeler-lcs [21.07.2017]

54. Macrologic (2017) Procesowy system ERP. Retrieved from https://www.macrologic.pl/erp/procesowy-erp/erp [1.07.2017]

55. Ultimus (2004) BPM—simplified a step-by-step guide to business process management with the Ultimus BPM Suite, Feb 2004

56. Ultimus (2004) Adaptive discovery. Accelerating the deployment and adaptation of automated business processes

57. Forrester (2016) The Forrester wave™: dynamic case management, Q1 2016. Published: 2 Feb 2016

58. Liu C, Li Q, Zhao X (2009) Challenges and opportunities in collaborative business process management: overview of recent advances and introduction to the special issue. Inf Syst Front 11(3):201–209

59. Kemsley S (2009) Dynamic BPM platforms. Retrieved from http://column2.com/2009/03/webinar-dynamic-bpm-platforms/ [3.04.2016]

60. Gong Y, Janssen M (2011) From policy implementation to business process management: principles for creating flexibility and agility. Gov Inf Q 29:61–71
61. Gartner (2017) Magic Quadrant for intelligent Business Process Management Suites. Technical Report ID: G00315642. Published: 24 Oct 2017
62. Gartner (2019) Magic Quadrant for intelligent Business Process Management Suites. Technical Report ID: G00345694. Published: 30 Jan 2019
63. Software AG (2015) Software AG—webMethods business process management platform. Retrieved from http://www.bpmleader.com/software-ag-webmethods-business-process-management-platform/ [3.04.2016]
64. ISIS Papyrus (2016) Adaptive processes. Retrieved from http://www.isis-papyrus.com/e15/pages/business-apps/adaptive-case-management/adaptive-process.html [4.04.2016]
65. IBM (2014) IBM business process manager V8.5.5 adds case handling and enhanced mobile UIs and IBM business monitor V8.5.5 provides more powerful analytics. Retrieved from https://www-01.ibm.com/common/ssi/cgi-bin/ssialias?infotype=an&subtype=ca&appname=gpateam&supplier=897&letternum=ENUS214-141 [20.07.2017]
66. Gartner (2016) Magic Quadrant for intelligent Business Process Management Suites. Technical Report ID: G00276892; Published: 18 Aug 2016
67. Gartner (2016) Magic Quadrant for BPM-platform-based case management frameworks. Technical Report ID: G00276724; Published: 24 Oct 2016
68. Rifkin J (2016) Społeczeństwo zerowych kosztów krańcowych. Internet przedmiotów, Ekonomia współdzielenia, Zmierzch kapitalizmu (The zero marginal cost society: the Internet of Things, the collaborative commons, and the eclipse of capitalism). Wydawnictwo Studio EMKA, Warszawa
69. Dwyer T (2013) Big Data and BPM. Retrieved from http://www.bpminstitute.org/community/bpm-group/big-data-and-bpm [9.07.2017]
70. Earls A (2013) What Big Data means to BPM: more event sensors, process simulations. Retrieved from http://data-informed.com/what-big-data-means-to-bpm-more-event-sensors-process-simulations/ [17.07.2017]
71. Pucher M (2012) How to link BPM governance and social collaboration through an adaptive paradigm. In: Fisher LW (ed) Social BPM: work, planning and collaboration under the impact of social technology. Future Strategies Inc., Lighthouse Point, pp 57–76
72. Forrester (2018) The Forrester wave™: Robotic Process Automation, Q2 2018. Published: 26 June 2018
73. Aalst W, Bichler M, Heinzl A (2018) Robotic process automation. BISE Retrieved from https://doi.org/10.1007/s12599-018-0542-4 [21.05.2018]
74. Flasiński M (2016) Introduction to artificial intelligence. Springer International, Switzerland
75. Pucher M, Ruhsam C, Kim T et al (2014) Towards a pattern recognition approach for transferring knowledge in ACM. In: 2014 IEEE 18th international enterprise distributed object computing conference workshops and demonstrations
76. Kim S, Hwang H, Suh E (2003) A process-based approach to knowledge-flow analysis: a case study of a manufacturing firm. Knowl Process Manage 10(4):260–276
77. Davenport T (2005) Thinking for a living. How to get better performance and results from knowledge workers. Harvard Business School Press, Boston
78. Marjanovic O, Freeze R (2012) Knowledge-intensive business process: deriving a sustainable competitive advantage through business process management and knowledge management integration. Knowl Process Manage 19(4):180–188
79. Kemsley S (2011). The changing nature of work: from structured to unstructured, from controlled to social. In: Lecture Notes in Computer Science Business Process Management, pp 2–2
80. Gartner (2016) Critical capabilities for BPM-platform-based case management frameworks. Technical Report ID: G00291102. Published: 3 Nov 2016

Chapter 3
Process Execution in a Knowledge Management Environment

Abstract Knowledge is the most important asset of the enterprise in the knowledge economy. Intellectual capital is fast becoming more important than financial capital. The key to success does not rest in amassing knowledge, but in its ongoing use and renewal in the business processes which create value for the organization and allow it to maintain competitive advantage. This requires the ongoing and consistent management of the entire knowledge within the organization—not just knowledge belonging to the management itself, but to the entire personnel. A particular challenge in this respect is the management of tacit knowledge, the existence which is often overlooked or hard to articulate, and which is naturally revealed and created in the course of work itself. For this reason, the aim of this chapter is to answer the following question: *Is it possible to integrate (dynamic) process management with knowledge management, including the management of tacit knowledge?* Section 3.2 discusses the term *knowledge* and presents two selected models of knowledge management in the organization, with a particular focus on the sources of knowledge, the awareness of lacking knowledge, and processes of knowledge renewal and verification. Organizations in the knowledge economy face the challenge of avoiding the risk of owning outdated knowledge and pseudo-managing such knowledge, as well as the challenge of maintaining the pace of acquiring knowledge and instituting its broad practical use. Sections 3.3 and 3.4 demonstrate how the expansion of traditional process management with the concept of dynamic process management enables organizations to create new knowledge on an ongoing basis, as well as implement mechanisms of constant knowledge verification. The ability to perform limited experiments and acquire knowledge in the course of business process execution allows for the continued creation of practical knowledge and its objective and independent verification on the part of the clients. The last section of this chapter 3.5, presents the consequences of integrating dynamic BPM with knowledge management with a view to creating a learning-by-doing organization.

Keywords Knowledge · Knowledge management (KM) · Explicit knowledge · Tacit knowledge · Knowledge codification · Knowledge acquisition · Dynamic business process management (dynamic BPM) · Case management (CM) · Process-oriented knowledge management (pKM) · Knowledge-intensive process management (kiBP) · Process mining

M. Szelągowski, *Dynamic Business Process Management in the Knowledge Economy*, Lecture Notes in Networks and Systems 71,
https://doi.org/10.1007/978-3-030-17141-4_3

3.1 Introduction

The significance of knowledge in the economy has been realized long ago. In *Principles of Economics,* published in 1890, Marshall [1] wrote: "Knowledge is our most powerful engine of production." However, economy would first have to mature enough to reach the stage in which information, knowledge, and their management became the fundamental source of competitive management alongside the pace of operations. There is no single commonly accepted definition of knowledge nor knowledge management [2]. For this reason, before we embark on an analysis of the requirements for knowledge management in dynamic business process management, we must establish what knowledge is and how processes of knowledge management in the organization are performed in the increasingly hypercompetitive environment—processes pertaining to revealing and transferring tacit knowledge in particular. The question, an answer to which may turn out to be decisive for the bottom line of many organizations, is as follows: *Are organizations able to manage tacit knowledge in a systemic fashion?* The aim of this chapter is to provide an answer to a broader and more practical question: *Is it possible to integrate (dynamic) business process management with knowledge management, including the management of tacit knowledge?*

3.2 Characteristics of Knowledge

Knowledge is a term which is challenging, if not downright impossible, to define in an unequivocal fashion. In relevant literature in the matter we can find multiple attempts at coining a definition, including ones by Nonaka and Takeuchi [3, pp. 48–72] and Jashapara [4]. One of the best-known of these attempts is a definition by Davenport and Prusak: "Knowledge is a fluid mix of framed experience, values, contextual information, and expert insight that provides a framework for evaluating and incorporating new experiences and information. It originates and is applied in the minds of knowers. In organizations, it often becomes embedded not only in documents or repositories but also in organizational routines, processes, practices, and norms" [5]. In consequence, knowledge should be understood as the fluid combination of information, experiences, and the context, which enable the ongoing evaluation and acquisition of new knowledge. In other words, knowledge is not, as Malhotra claims, "organized information" [6, pp. 5–16], as information in accordance with its definition must contain organized data. Knowledge in and of itself cannot define the goals of acting or refraining from taking action. It can be used for smart or foolish reasons, ethical or malevolent reasons, good or bad reasons [7, pp. 46–48].

For the purposes of this work, we will make use of a narrower definition of knowledge, which is closer to the definition of specialist knowledge [8, 373]:

Knowledge is a skill (individual or collective) resulting from experience, which allows for the evaluation, interpretation, and use of information in the specific context of its acquisition.

From an infological perspective, we can define knowledge as a function of information, context, and experience:

$$\omega = <\mathbf{I, C, E}>$$

where

(**I**) Information stands for data[1] organized through e.g. categorization and classification; processed with a view to providing a systematic and structured form allowing for the quantitative analysis of phenomena [9, p. 96].
(**C**) Context stands for the circumstances, surroundings, background (e.g. cultural), or ties of the specific event in question (an experience or information acquisition). It is the context which enables the proper interpretation of the information or the experience.
(**E**) Experience is based on events (or their sequences—processes) which have taken place in the past and which have left a mark in the mind of the individual or in collective memory. They arise on the foundation of personal views, sensibilities, mental attitudes, or even biases and proclivities. Experiences can be treated as the awareness of that which is, or the skill of foreseeing the effects which might arise in circumstances resulting from information acquired in a given context [10].

Often, other aspects of knowledge are also looked upon, such as:

- intuition;
- judgment skills;
- values;
- assumptions;
- views;
- held objective and subjective truths.

Knowledge is inextricably tied to experience. It is shared not only through communicating data or information (science), but also through personal experience or the analysis of the actions of others. Knowledge combines the acquired information with experiences made as the result of trial and error. For example, even individuals who ride a bike daily often find themselves unable to pass on this skill. In turn, individuals who do not know how to ride a bike cannot learn this skill from a handbook on bike-riding. However, if we were to attempt to write a handbook for beginner bike-riders or hold a training session with a person who is just learning to ride, we could nevertheless become aware of many issues that we did not notice on a daily basis. By helping someone understand how to maintain their balance or how to best perform specific maneuvers, we are able to notice better, previously overlooked methods of performance. Beside showing the role of experience and the context of sharing knowledge (riding a bike), the above example also showcases four essential attributes of knowledge, which clearly distinguish knowledge from tangible resources.

---

[1]For the purposes of this work it was assumed that data are phenomena represented by unconnected facts, that is, past events of an unprocessed nature which then act as a source of data.

Simultaneity

"For unlike land and machines, which can be used by only one person or firm at a time, the same knowledge can be applied by many different users at the same time" [9].

Inexhaustibility

Knowledge is best multiplied by being shared! Knowledge grows, not shrinks, with use. In this way, it is easy to broaden the scope of data collection, data analysis, and intuition, as well as to accelerate the pace of collecting experiences.

Impermanence

Knowledge depreciates with time: it loses value as it becomes outdated. Outdated knowledge may lead to losses, which de facto means it has negative value.[2]

Nonlinearity

There is no explicit correlation between the size of one's knowledge resources and the results thereof, including benefits. A large body of knowledge may not necessarily provide huge results. And contrariwise: a small body of knowledge may have disproportionately large effects.

Fundamental practical problems pertaining to knowledge management include:

Dispersion of information as a component of knowledge

(increasingly more often: information overload)

Information is found in many places and in many forms. It is not always apparent where a given piece of information may be found and which of its versions is up to date [11, pp. 3–4].

Relations between information

It is not always apparent just how specific pieces of information are connected to one another with a view to forming a body of practical knowledge. It is often challenging to find among multiple pieces of information the one which is responsible for failure or success.

Suboptimization

One recurring problem in the data and information entry stage is the distortion, or suboptimization, of knowledge to conform to the particular interests of organizational units or specific individuals.

Codification

Another challenge rests in convincing the employees to systematically acquire and codify data and information. This step is often perceived as an additional, burdensome requirement or even part of a system of personnel invigilation on the part of the management. In the traditional approach, the codification of knowledge is seen as a cost-intensive and inefficient process [12].

Timeliness

A recurring problem rests in verifying whether knowledge is up to date, as there are risks involved in using outdated knowledge.

---

[2]The pace of devaluation of knowledge may differ depending on the type of knowledge in question. A significant devaluation in value of knowledge pertaining e.g. to IT or pharmaceutics is a matter of just several years, while knowledge pertaining to e.g. the principles of human anatomy or geography remains up to date (Earth, after all, is still round).

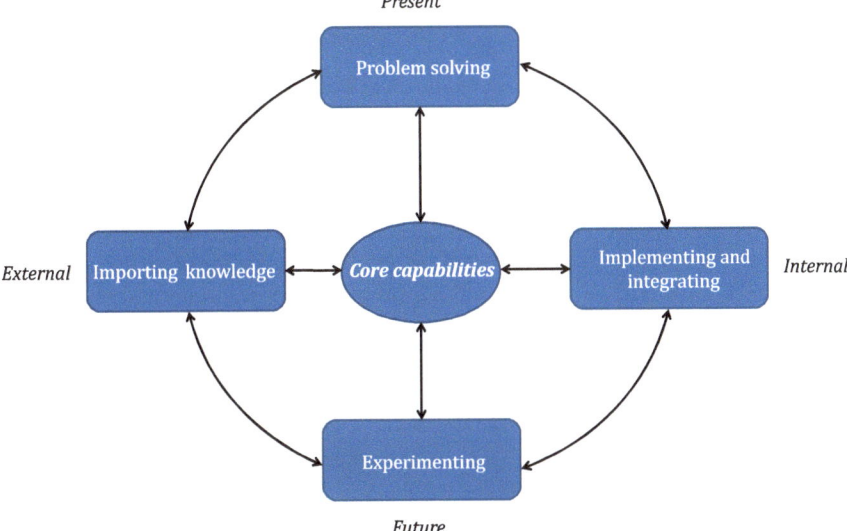

Fig. 3.1 Dorothy Leonard's model of the wellsprings of knowledge. *Source* Author's own elaboration on the basis of Leonard [13]

Sources of knowledge

A fundamental practical problem tied to knowledge dispersion, suboptimization, and codification is the problem of the sources of knowledge. In accordance with Dorothy Leonard's model of wellsprings of knowledge presented in Fig. 3.1 [13, pp. 8–12], efficient management of knowledge requires the existence of five components (skills):

- the organization's interior—implementation and integration of new tools and solutions;
- the organization's exterior—knowledge acquisition from the environment;
- ongoing operations—creative problem-solving, dissemination of knowledge and mutual search for the most efficient solutions;
- development efforts—experimentation, formal and informal search for better, more innovative solutions.

The fifth component combines all these elements into a single, efficient system comprises core capabilities, which include the skills and knowledge of the personnel, management systems, technical systems, as well as binding norms and values.

3.2.1 Tacit and Explicit Knowledge

We are able to distinguish between two types of knowledge:

- **Explicit knowledge**, sometimes referred to as formal, available, or expressible knowledge, is characterized by the formalized acquisition of information, the description of the context of its use, as well as guidelines on practical experience—usually derived from the available information ("I know «what»") and the context of its acquisition. In organizations, explicit knowledge often takes the form of codified knowledge available in the form of procedures, rules and regulations, quality standards, process models, databases, methodologies of management, technologies, descriptions of production processes, instruction films, etc., available in accordance with one's level of privileges and updated in accordance with principles set by the management of the organization.
- **Tacit knowledge**, sometimes referred to as confidential, silent, or quiet knowledge, is often overlooked in organizations. It is owned by particular individuals or teams of individuals. It is derived from experience ("I know «how»"), being brought up in a particular cultural network, innate skills, etc. Within teams, this knowledge manifests itself in the organizational culture, views and attitudes, cultural norms, means of interpersonal communication, etc. This type of knowledge is usually revealed and developed in specific contexts of operation, outside of which access thereto is more challenging or even outright impossible. This knowledge also contains intuitions, assumptions, and subjective judgments, the sources of which are often not logically and formally documented, or even expressed by their owners—individuals or teams. However, this in itself does not prevent the owners of this knowledge from acting upon it, verifying it, or sharing it.

Table 3.1 summarizes the differences between tacit and explicit knowledge.

Both types of knowledge are not exclusive or independent of one another [16]. They comprise different aspects of a changing continuum. Explicit knowledge is one in the case of which the element of personal experience, as well as often the element of context is limited, and information itself is clearly foregrounded. Nevertheless, interpreting and understanding such knowledge requires both an awareness of the context and the necessary experience, even if it is limited to simply being brought up in a given culture. In turn, tacit knowledge is one whose main element is experience. However, this experience nevertheless is derived in a specific context and requires in the least the bare minimum of information (Fig. 3.2).

The simplified division enforced by the strict tacit/explicit knowledge binary is, in effect, artificial [17]. All knowledge must contain all three components (information, context, and experience) or it will not be knowledge as such, but just information (a common mistake is to refer to repositories of information as "knowledge bases").

The Availability of Tacit Knowledge

Tacit knowledge is characterized by its limited availability to potential recipients. It is estimated that knowledge which may be codified and formally transferred comprises

Table 3.1 Differences between tacit knowledge and explicit knowledge

	Tacit knowledge	Explicit knowledge
Main features	Personal	Impersonal
	Subjective	Objective
	Contextual	Independent of context
	Hard to document	Documented and codified
	Hard to disseminate within the organization	Simple to share
Examples	Expertise	Knowledge repository
	Routine team operations	Handbooks, case studies
	Informal business processes	Rules and regulations, procedures, and processes
Benefits	Hard to copy or steal	Simple to share and transfer
	Rich in details and nuances	Simple to manage
	Created and updated in the course of practical operations	Measurable
	Belongs to the person	Belongs to the organization
	Strong ties with organizational and social culture	Simple to provide an objective cash-flow valuation
	Tied to specific individuals	Patentable and sellable
Drawbacks	Transferred between individuals during operations	Must be constantly verified and updated
	Hard to identify and codify	Requires adaptation to new contexts
	Hard to manage	Requires familiarization and practical confirmation
	Not owned by the company	Easy to copy
	Cannot be patented and sold	"leaky" and easily stolen
	Lost if the person leaves or dies	Hard to verify, especially when there are too many sources of knowledge

Source Author's own elaboration on the basis of Tiwana [8, p. 65], Hislop [14, pp. 15–20], Talisayon and Talisayon [15]

"I know HOW"	**"I know WHAT"**
Inteligence Focus on operations Being capable of performing tasks	Owning knowledge Container metaphor "Being"

The knowledge continuum

Fig. 3.2 The knowledge continuum according to Gilbert Ryle and Michael Polanyi. *Source* Author's own elaboration, on the basis of Jashapara [4, p. 60]

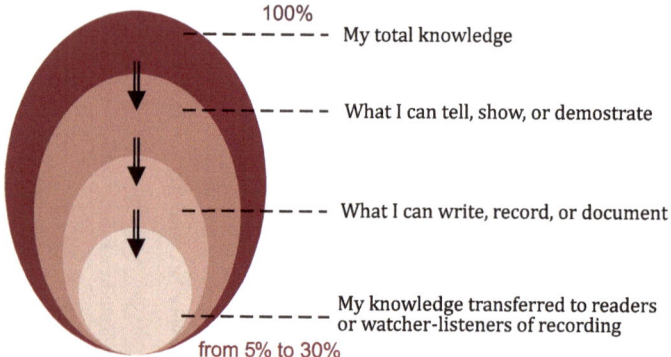

Fig. 3.3 Losses of knowledge in the process of its revealing (externalization). *Source* Author's own elaboration, on the basis of Talisayon and Talisayon [15]

just between 5 and 30% of the entire body of knowledge. Figure 3.3 shows in a concise manner the losses of knowledge in the process of its revealing (externalization).

Losses of knowledge in the process of its revealing are the result of the fact that individuals owning knowledge:

- have more knowledge than they are able to transfer in dialogue with others;
- in the course of dialogue or presentation are able to share (in verbal form, but also through intonation, facial expressions, and gestures) more than they are able to share in writing or document in another permanent form.

Due to the fact that tacit knowledge is strongly tied to specific individuals, it is often taken for expert knowledge—due to the fact that it is contained in the minds of experts [18]. This is the result of difficulties with (self-)expressing or even verbalizing knowledge by its owners. The second reason rests in the fact that owners of tacit knowledge are often unaware of having the knowledge in the first place ("do not know that they know") [7]. This may be the result of the lack of the ability to articulate knowledge, such as e.g. limited vocabulary, limited conceptual apparatus, or lack of a formal synthesis between different pieces of information and experiences. This particularly pertains to practitioners, who are freely able to make use of their skills in action, but who nonetheless have problems with verbalization or organized transfer.

Usually a much larger challenge rests in convincing an expert or a team of experts to take the effort to share and codify their knowledge. Even when experts have access to the necessary vocabulary, they may nonetheless face challenges with respect to describing the full context of using, or the usefulness of, knowledge. Such limitations result in experts attempting to share their knowledge only following strong organizational pressure. In effect, knowledge is not being shared in the course of the experts' ongoing actions under normal working conditions, but is instead shared in the form of one-off actions or codification projects. Knowledge shared in this way is incomplete and often devoid of the full context enabling its proper implementation. At the same time—due to the rapid pace of changes—knowledge revealed in this way often becomes outdated after a short time and requires constant updating and renewal. Actions made on the basis of outdated knowledge may cause measurable losses to the organization [19].

Tacit knowledge usually does not have the form of personal knowledge belonging to a single individual. Most projects and processes are not able to be performed by a single individual, even if that individual is a world-class expert. They naturally require cooperation between teams and the harmonious combination of individual expert knowledge and social skills, which are manifested only in the course of teamwork. For example, a world-class musician with undoubtedly unique personal professional skills may not be the best fit for an orchestra. By analogy, a soccer player may have great technique, but fail as a team player. In other cases, the opposite is true: average, mediocre individuals are able to reach great heights as part of a larger team. A good team has unique collective knowledge resulting from the combination of individual knowledge and social skills, which are usually a part of the unrecognized tacit knowledge (synergy effect).

The Creation of Tacit Knowledge

Determining the sources of tacit knowledge is a challenging task. Figure 3.4 shows the components behind the creation of tacit knowledge (internalization).

Learning-by-doing is a much more effective source of tacit knowledge than academic knowledge, as in this process new knowledge is created through self-

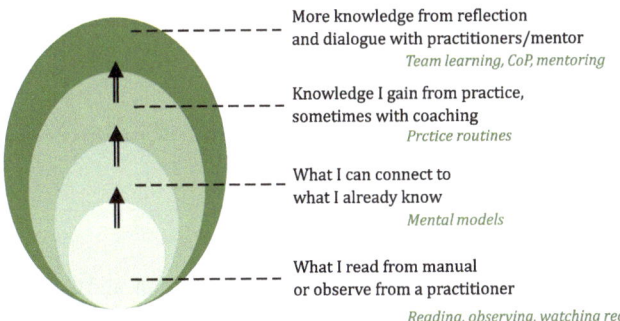

Fig. 3.4 Components of the process of acquiring tacit knowledge (internalization). *Source* Author's own elaboration, on the basis of Talisayon and Talisayon [15]

improvement, incremental innovations, limited experiments—and immediately integrated with the existing body of knowledge [20]. Other authors point to the significance of "inaccessible" memory [21]. Declarative memory stores knowledge which may take the form of a conscious, planned-out statement, e.g. "When I was performing surgery on a patient with the use of procedure X, it has led to result Y." "Inaccessible" memory offers no such possibilities, which means it is closer to the concept of tacit knowledge. Tacit knowledge is naturally the easiest to uncover in action, e.g. in the course of performing work. This, however, requires the creation of organizational conditions in which employees are formally empowered to decide on the method of performing their work (in accordance with the 2nd principle of dynamic BPM), as well as required to document it (in accordance with the 3rd principle of dynamic BPM) [22]. In effect, in accordance with the 3rd principle of dynamic BPM, it is possible to capture not just fundamental information (e.g. the decisions made and the key factors behind them), but also the entire context of acquiring or verifying knowledge, including e.g. the skills of the process performers or the factors with second- or third-degree significance. In this way, this "unrealized knowledge," or "private knowledge," which was the property of individuals or teams of individuals who created and used the knowledge in question, may be transformed into explicit knowledge. The organization may gain insight into the full context of its creation, verification, or discovery, including access to the employees who have it. This considerably changes the significance of the term "expert" within the organization. From individuals who are detached from daily operations and hidden in research and development departments they may simply become "knowledge workers": workers using and creating their (tacit and explicit) knowledge in the course of daily work. Knowledge acquired and updated in this way is practical in nature and available to be used in diagrams or checklists of performed processes, justifications of or commentaries on introduced process changes, or detailed analyses pertaining to the use and the results of using specific process patterns. It is not created in the course of some one-off knowledge-sharing "actions" or "projects," organized for the sake of creating knowledge alone. It is the result of solving specific problems, and so it can be easily analyzed, as well as codified on the basis of systems or taxonomies closely aligned with the genuine needs or the processes of the organization and its clients [23].

3.2.2 Processes Pertaining to the Creation and Management of Organizational Knowledge

Knowledge management is:

- the totality of processes enabling the creation, dissemination, and use of knowledge with a view to fulfilling the goals of the organization [4, 24][3];

---

[3]In 1997, Peter Murray and Andrew Myers from Cranfield School of Management in Great Britain have conducted wide-scale research, during which they presented management of different enter-

- according to Dorothy Yu from PricewaterhouseCoopers—"The art of transforming intellectual assets into business value" [25, p. 61];
- according to Wiig—"The systematic and explicit management of knowledge-related activities with the goal to build and exploit intellectual capital effectively and gainfully" [26, p. 3].

The operative phrase in Wiig's definition is "systematic." Due to ongoing qualitative changes described in Chap. 1, knowledge which is not kept up to date in many instances quickly becomes outdated. The faster the pace of changes to the surrounding environment, the faster the process. In effect, it is crucial to properly manage:

- the creation or acquisition of new knowledge;
- sharing knowledge;
- using, developing, evaluating, and verifying knowledge;
- renewing owned knowledge or discarding outdated and useless knowledge;
- the awareness of lacking knowledge [27, p. 163, 28, p. 86].

This is a process without a clear-cut, perfect model, which would fully describe all actions and interactions. For this reason, scholarship in this regard points to multiple models of knowledge management [3, pp. 111–117, 29, p. 46]. In order to briefly describe the key areas of knowledge management, we will make use of Nonaka's and Takeuchi's model and the process-oriented knowledge management model.

3.2.2.1 Nonaka's and Takeuchi's Knowledge Management Model

According to Nonaka and Takeuchi [3], fundamental significance to using and creating knowledge rests in interactions between tacit and explicit knowledge, which allow for the conversion of knowledge in four different manners:

1. (S) Socialization—from tacit knowledge to tacit knowledge,
2. (E) Externalization—from tacit knowledge to explicit knowledge,
3. (C) Combination—from explicit knowledge to explicit knowledge,
4. (I) Internalization—from explicit knowledge to tacit knowledge.

The aforementioned interactions create a spiral of managing individual or collective knowledge, within which, on an ongoing basis, new tacit knowledge is first updated, and then created, after which it is revealed and combined with other available components of explicit knowledge, internalized (familiarized) as individual or collective tacit knowledge updating the available knowledge and forming the basis for the creation of new knowledge, etc. (Fig. 3.5). The spiral is executed in the course of a five-phase model of the organisational knowledge creation process (Fig. 3.6).

---

prises with a collection of definitions pertaining to knowledge management, asking the management to select the definition which in their mind best reflects the essence of knowledge management. As many as 73% of the respondents pointed to the definition cited above.

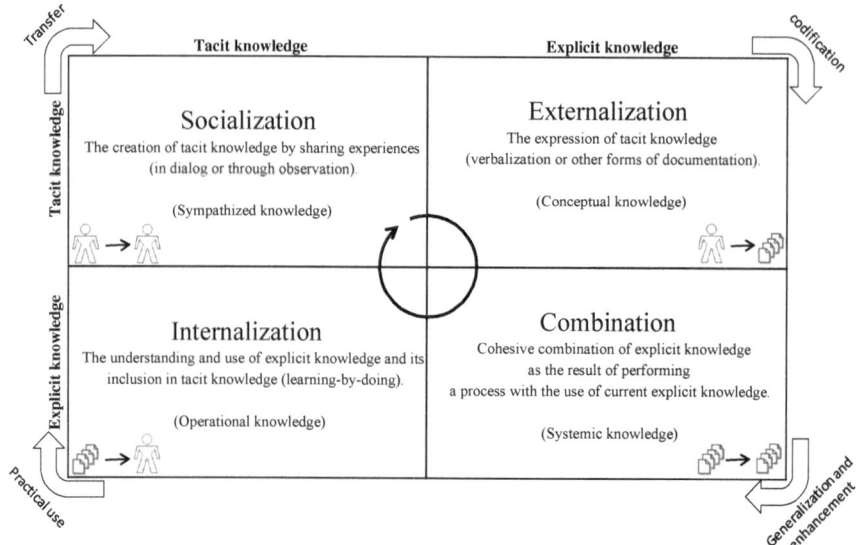

Fig. 3.5 Knowledge management spiral (SECI model). *Source* Author's own elaboration, on the basis of Nonaka and Takeuchi [3]

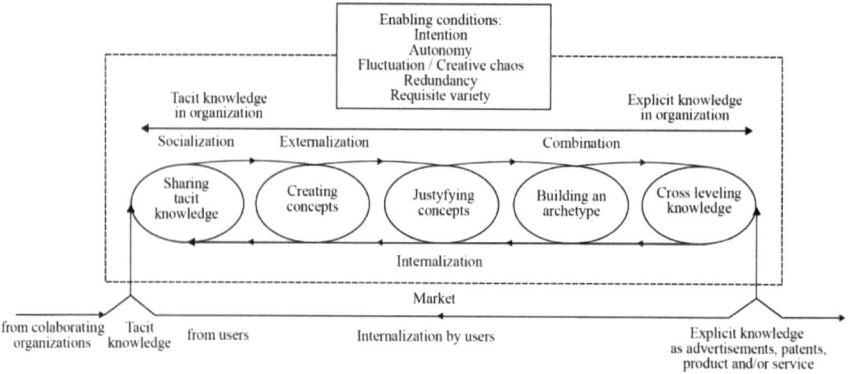

Fig. 3.6 Five-phase model of the organisational knowledge creation process. *Source* Nonaka and Takeuchi [3]

The model assumes that specific phases (which de facto are subprocesses) of the organizational knowledge creation process have the following significance and encompass the following actions:

1. Sharing tacit knowledge (corresponds to knowledge socialization)
   The process begins with sharing personal tacit knowledge within teams. It creates a community of modes of thinking, mental models, and patterns of action between the members of the team.
2. Creating concepts (corresponds to knowledge externalization)

This is the most intensive phase of the process, in which tacit knowledge is expressed with a view to creating explicit knowledge in the form of an idea or concept by way of its verbalization or in some other form (e.g. a model, drawing, video, example of action). Such creative ideas are formed in response to the needs of the organization, tasks set before it, observed or anticipated requirements of the market, the competition, or the clients. One natural source of ideas are the actions of the employees or their teams, which are undertaken in the course of performing work in accordance with the principle "necessity is the mother of invention." In the course of internal team discussions (dialogue, brainstorming, work on a model, formal development of improvement recommendations or strategic scenarios, etc.) they ultimately take an explicit form, which is readily available for critical discussion and verification. They usually do not have the form of optimal or ideal solutions, but instead of heuristics enabling the performance of tasks set before individual employees or teams of employees, with the option of further elaboration and optimization. In this phase, premises for experiments are also defined with a view to verifying the validity of the new ideas and concepts.

3. Justifying concepts
   Because for Nonaka knowledge is a "confirmed belief" [3], ideas which were created and not dismissed in the previous phase should be verified in practice. This repeatedly necessitates the formulation of new explicit knowledge in the scope of the formalized requirements and criteria of verification, encompassing e.g. market requirements or requirements resulting from the organization's strategy, as well as fundamental criteria of evaluation on the part of the clients, the management, or other stakeholders. The result of the verification usually takes the form of an idea or a concept of improvement, which is better suited for communication within the organization.

4. Building an archetype
   This phase consists of combining new knowledge with existing explicit knowledge. In production companies this phase corresponds to building a prototype, whereas in service companies—to defining good business practices. In this phase, the concept (idea) takes shape and becomes ready for broad dissemination, usually by way of changes to previous versions of the prototypes or good practices in question. At the same time, in this phase negative approaches are identified with a view to creating a database of wrong practices, despite the fact that some of these wrong practices may have been even recommended in the past.

5. Cross-leveling knowledge (corresponds to knowledge combination)
   In this phase, new or updated knowledge is shared and acquired in practice. The knowledge usually takes the form of changes to services, products, technologies, business practices, or business processes—both within the organization (individually or collectively) and between organizations (in a virtual or network organization, or in a chain of organizations which create value for the client, etc.). It is predicated upon changing mental models and behavior patterns by understanding individual or collective experiences on the basis of the new or updated knowledge made available. Information, knowledge, and skills should no longer be closely tied to specific actions, organizational silos, or even specific individ-

uals or locations in which the organization operates. In result, it is possible to avoid wasting great amounts of time and work on acquiring knowledge which is in fact available within arm's reach, and the organization itself is capable of using this acquired knowledge without repeating the same mistakes time and again.

Cross-leveling knowledge is a process which triggers another, new organizational cycle: the creation of knowledge in the form of a feedback loop initiated by the interaction between employees (or their teams) with access to tacit knowledge enriched in the cross-leveling knowledge phase. This further dissemination of knowledge initiates the search for, as well as the confirmation of, new ideas based on this new knowledge, which are then used to build a pattern which is shared and assimilated (internalized) as tacit knowledge in the course of the cross-leveling process—after which the entire cycle begins anew. The internalization of knowledge should immediately be accompanied by the broad collection of experiences with a view to triggering further ideas and exploratory actions. In order to make this possible, it is crucial to empower individuals and teams with the necessary autonomy to freely draw upon knowledge created elsewhere and use it adequately to their needs. Concepts formulated in one cycle should trigger another cycle of improvement on a higher level. The knowledge management process should never end, but instead consequently ascend the organization to a higher level—by improving the quality and accessibility of its body of knowledge. This should result in the day-to-day search for innovations with a view to raising the efficiency of operations and providing new competitive advantages in order to retain old clients and win new clients (from the competition).

One fundamental risk which should be kept in mind at all times is the threat of owning and pseudo-managing outdated knowledge. In order to counteract this threat, the knowledge collection mechanism should be supplemented with a mechanism of discarding (unlearning) outdated knowledge. Knowledge management should create and institutionalize the organization's capability to reinvent and reform itself. The key to success lies not projecting the future, but in constantly changing the principles of operation in an unforeseeable and surprising future. This requires organizations to constantly discover the best possible means of operation with a view to satisfying their stakeholders (clients, owners, employees, etc.).

3.2.2.2 The Process-Oriented Knowledge Management Model

The process-oriented knowledge management model is based on practical solutions. It has been developed on the basis of the experiences of large organizations, primarily large consulting companies. In principle, the process-oriented knowledge management model is based on the resources and processes which are already in existence in the organization, including existing organizational culture, leadership, and infrastructure. Davenport and Prusak [5] distinguish between three knowledge management processes, which influence one another by way of feedback loops:

- The knowledge acquisition process (often referred to as the knowledge creation process)

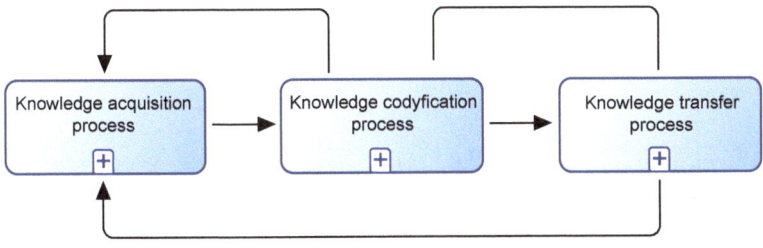

Searching for information and knowledge	Revealing knowledge	Distributing information and knowledge
Selecting information and knowledge	Gathering knowledge	Transferring knowledge in the course of formal operations
Creating and updating knowledge	Codifying knowledge	Transferring knowledge in the course of informal operations
Acquiring new knowledge	Preparing knowledge for distribution	

Fig. 3.7 The process-oriented knowledge management model. *Source* Author's own elaboration, on the basis of Davenport and Prusak [5]

- is a process with a view to raising the quantity of knowledge owned by the organization by way of acquiring knowledge from the outside or creating it internally. Efficient performance of this process provides the organization with the ongoing capability of acquiring new knowledge;

- The knowledge codification process

  - is a process of revealing and gathering knowledge, as well as preparing it for dissemination, with a view to presenting knowledge in such a way, as to make it accessible to all members of the organization. In result of this process, knowledge acquired in the organization is explicit and available in a clear, accessible form;

- The knowledge transfer process

  - encompasses knowledge dissemination and acquisition. In the case of explicit knowledge the transfer is usually performed with the use of ICT systems and tools (IT systems, decision support systems, data warehouses, social media apps, workgroups, the organization's intranet, audio- and video-conferences, etc.). In the case of tacit knowledge, the transfer is performed by way of direct formal or informal contacts between individuals and teams (e.g. in the course of performing work, training workshops, or employee development).

The process-oriented knowledge management model presented in Fig. 3.7 in Davenport's and Prusak's authorial version [5] does not single out the process of using knowledge with a view to making decisions or performing work. For this reason, multiple authors postulated its supplementation with a knowledge use process, which would encompass knowledge acquisition and the transformation of knowledge into decisions [30].

3.2.3   The Most Common Problems Pertaining to Knowledge Management

It is possible that in the course of implementing the knowledge management process in the organization certain problems will arise or certain mistakes will be made in consequence of misunderstanding the essence of knowledge and underestimating the significance of the cultural components of implementation. The most common of such errors in judgment, which were indicated as far back as in 1998 by Prusak and Fahey [31], are presented in Table 3.2.

After almost two decades since its inception, it would seem that implementations of knowledge management have the largest consequences with respect to errors in terms of:

- overstating the significance of technology and ICT tools and underestimating the significance of the knowledge workers themselves:

  - error 2: Emphasizing knowledge stock to the detriment of knowledge flow
  - error 3: Viewing knowledge as existing predominantly outside the heads of individuals
  - error 5: Paying little attention to the role and importance of tacit knowledge
  - error 10: Substituting technological contact for human interface

This is often the result of a lack of awareness of the difference between information and data, as well as treating explicit knowledge as the total body of knowledge available to the organization. However—as has been demonstrated in Sect. 3.2.1—even following the introduction and execution of the requirement of codifying knowledge available to the employees, part of the tacit knowledge will not be revealed and codified regardless;

- disentangling knowledge management implementations from the day-to-day operations of the organization (the performance of the fundamental process):

  - error 6: Disentangling knowledge from its uses
  - error 5: Paying little attention to the role and importance of tacit knowledge
  - error 9: Failing to recognize the importance of experimentation

Knowledge created or acquired by sectioned-off teams (e.g. research and development departments) is often transferred to the knowledge workers in the form of an "algorithm to execute," without a more detailed analysis of the actual possibilities offered by its practical use or its harmonious alignment with the organizational culture. This rids the organization of the capability to verify the practical value of the new knowledge on an ongoing basis. Within traditional business process management, which does not offer the possibility of limited experimentation, it enables

Table 3.2 *Eleven deadliest sins of knowledge management*—typical errors of implementation projects

Error 1: Not developing a working definition of knowledge	• Problems with distinguishing between knowledge, information, and data • Lack of reflection on the goals of knowledge management initiatives
Error 2: Emphasizing knowledge stock to the detriment of knowledge flow	• Knowledge codification instead of knowledge personalization • Gathering, archiving, measuring—instead of sharing or using knowledge
Error 3: Viewing knowledge as existing predominantly outside the heads of individuals	• Gathering knowledge resources controlled by the organization • Underestimating the role of the employees in managing knowledge
Error 4: Not understanding that a fundamental intermediate purpose of managing knowledge is to create shared context	• No incentives to share knowledge • Improper work atmosphere, limited trust, personal conflicts
Error 5: Paying little attention to the role and importance of tacit knowledge	• Underestimating the experiences of the employees and the significance of knowledge personalization
Error 6: Disentangling knowledge from its uses	• The initiative of knowledge management becomes an end onto itself: the construction of an archive, measurement of the value of intellectual capital, market research • Forgetting that knowledge should be used in practice
Error 7: Downplaying thinking and reasoning	• Only an analysis of owned knowledge resources will allow us to make better decisions
Error 8: Focusing on the past and the present and not the future	• Past experiences do not necessarily help us act in new situations
Error 9: Failing to recognize the importance of experimentation	• The initiative of knowledge management as an independent, separate program—not as a collection of multiple techniques introduced by way of trial and error
Error 10: Substituting technological contact for human interface	• The belief that an IT system will solve all of the problems connected with knowledge management
Error 11: Seeking to develop direct measures of knowledge	• Measuring knowledge on the basis of the number of documents, searches, training workshops, and implemented initiatives • Forgetting that the most important indicators pertain to tangible effects of using knowledge

Source Author's own elaboration, on the basis of Fahey and Prusak [31, pp. 265–266]

the organization to tailor knowledge to its needs or improve upon it further. In effect, devoid of dynamic business process management, knowledge management becomes counter-efficient and may even, against all expectations, diminish the competitive advantage of the organization. It should be note that without the concurrent implementation of process management and knowledge management (process-oriented KM—pKM), implementing dynamic business process management alone or using process exploration tools is to a large degree a futile undertaking. Its benefits would be thwarted regardless, as the knowledge created with the help of dynamic BPM would not be properly and quickly identified, codified, and transferred.

The next two decades since the publication of *Eleven deadliest sins of knowledge management* have seen the identification of further fundamental risks, threats, and errors which might cause seemingly successful implementations of knowledge management to deviate from their goals. The most important of them is information overload, which is tied to the wide availability of ICT solutions and their daily use in the social culture [11].

The sheer codification and collection of information and knowledge resources may not be conclusive to making more informed decisions due to information overload. The sheer quantity of information available in the Internet age which need to be taken into consideration may result in a phenomenon termed "paralysis by analysis," result in the knowledge workers feeling lost and demotivated. In response to this threat, some researchers go as far as to propose the management of ignorance, or the awareness of having incomplete knowledge [28]. Regardless of the degree of complexity of the procedures of knowledge codification and the ease of use of the systems in which the knowledge is stored—the largest contribution to knowledge management in organizations is made by the employees themselves. Within knowledge management initiatives, one should not forget about the benefits of personalization strategies based on communication and cooperation, which stimulate the transfer of tacit knowledge and facilitate reaching organizational experts. Procedures and systems will prove ineffective in organizations with a culture which tends to make employees gravitate toward hiding their knowledge instead of sharing it. Trust, a conductive work atmosphere, and properly selected motivators are a necessary condition of transforming the organization into an knowledge-oriented enterprise.

3.3 Mutual Relations Between Knowledge Management and Dynamic Business Process Management

The nature of knowledge differs significantly than the nature of tangible assets, e.g. means of production or financial capital. However, managers in organizations often strive toward extending strict control over knowledge management—analogous to their control over financial management. In effect, this leads to a situation in which companies forget that the fundamental capital in the knowledge economy is not financial capital, but intellectual capital. Because unused capital leads to losses, organizations which are governed on the basis of the knowledge of their managers alone in fact commit mismanagement, which radically lowers their chances of success on the market. What chances in an industrial economy would have an organization which used just a fraction of its capital? What chances, then, does an organization have in the knowledge economy when it limits itself to knowledge available to its management and:

- uses a small fraction of its total available knowledge;
- does not use knowledge, and hence knowledge is not multiplied;
- does not renew or verify knowledge, so it suffers losses resulting from having no access to current knowledge or using outdated knowledge?

It is, of course, possible to:

- acquire knowledge from outside the organization—but this course of action is expensive and we can never be sure whether or not the obtained knowledge has signs of aging—it might not provide the organization with competitive advantage, as it had been available to the competition for some time;
- create a research and development department in the organization—though practice shows that such units often become detached from market requirements.

Both of the abovementioned paths of acquiring knowledge should be used, but in the knowledge economy their use alone is insufficient. The goal set before organizations is the ongoing, constant acquisition of knowledge on process execution and its rapid use in process improvement on an organizational and extra-organizational level. This is only possible by the harmonious combination of knowledge management and (dynamic) process management. Only then is it possible to acquire and verify knowledge derived from all potential sources, including tacit knowledge accessed in the course of creative problem-solving and experimentation, following which knowledge is used and shared.

3.3.1 Managing Tacit Knowledge Within Dynamic Business Process Management

Since the nineties, business process management and knowledge management are in constant development, which is both increasingly strengthened by, and itself strengthens in return, the use of ICT technologies in management. This mutual relation between process management and knowledge management has been acknowledge for a long time. Certain early proponents of knowledge management, such as Davenport, have significantly contributed to the development of process management as well [32]. This trend has been slowed down by the rapid development of process management in the direction of business process reengineering. Its aspiration to blindly reject the business models owned by the organization, as well as its focus on

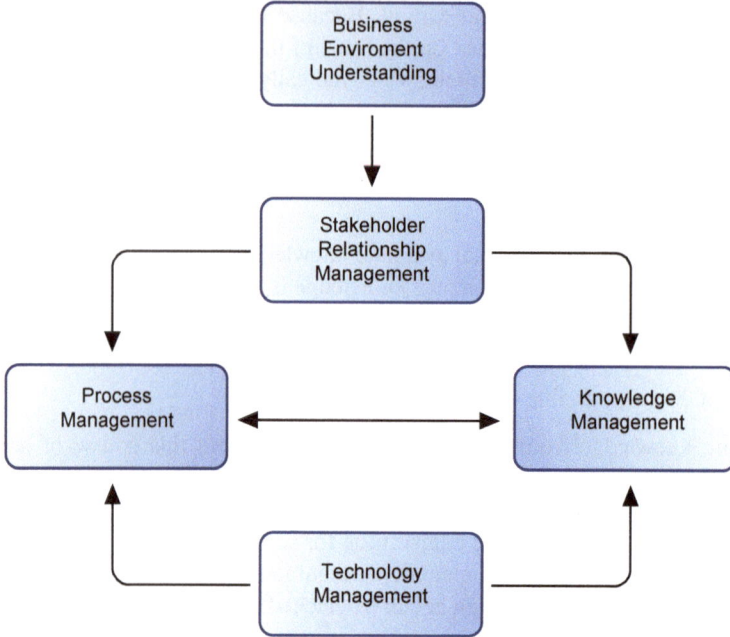

Fig. 3.8 Mutual relations between critical fields in raising business efficiency. *Source* Burlton [34]

restructuring, which in practical terms was aimed at downsizing, often lead to large-scale losses of knowledge and the implementation of practices which ran counter to the principles of a learning organization [27, 33]. Only after a decisive shift away from business process reengineering due to multiple failures which mostly resulted from the failure to account for the cultural dimension and great losses in knowledge did organizations return to attempts to research and use the synergy offered by process management and knowledge management. As shown in Fig. 3.8, as far back as in 2001 Burlton [34, p. 6] has conceptualized process management and knowledge management as fundamental dimensions of raising business efficiency. He has also stressed the need for their strict, mutual dependency.

With time, increasing attention has been paid to the benefits derived from including issues pertaining to knowledge management with a view to improving business process management [35]. In effect, specific methodologies were proposed [36–39]. However, such attempts were not known and used widely enough to significantly influence the direction of the development of business process management. Their primary flaw rested in the unsolved problem of the sources of knowledge within the organization. The concept of process-oriented knowledge management (pKM) has introduced the term knowledge-intensive business processes (kiBPs), but they were defined more as processes which intensively use knowledge acquired in the identification stage or in subsequent cycles of process improvement rather than processes which create or verify knowledge in the course of performance itself. Due to the limitations imposed by traditional business process management, process execution may only provide knowledge on its efficiency, productivity, used or involved resources, etc. After all, they do not offer process performers the possibility to shape or adapt processes to the context of the specific performance, which means that in effect they do not offer the possibility of creating new knowledge or revealing the tacit knowledge owned by process performers.

One could think that the integration of knowledge management and process management should have been halted for good due to the abovementioned limitations of both concepts. However, at the turn of the 20th and 21st centuries in effect of the growing pressure of user requirements extensions of BPMS systems started to appear, which allowed for modeling processes in the course of performance itself [40–43]. As has been discussed in Chap. 2, that time period witnessed the almost parallel development of methodologies and tools dedicated to the automated business process discovery (APBD), which have then been generalized as process mining methodologies. They allowed for the identification of knowledge generated in the course of performance itself [44, 45]. In 2012, which saw the publication of the Process Mining Manifesto, in the works of e.g. Marjanovic and Freeze it is clearly stated that knowledge is an integral part of business process management [46, 47]. In accordance with the principles of dynamic business process management (the 2nd principle dynamic BPM), this requires organizations to subordinate the process-based (procedural) component to the human component, and not vice versa, as it had been before. This considerably changes the significance of the term "knowledge-intensive processes," defining them as processes which use, verify, create, and reveal knowledge [46, p. 183].[4]

Management in the knowledge economy requires the use of the entire available knowledge, both explicit and tacit in nature. As has been discussed in Sect. 2.1, tacit knowledge is inaccessible in nature, which means that it is hard to approach and codify in the form of e.g. procedures, rules and regulations, or process descriptions. In the classic understanding, tacit knowledge is contained only in the mind of its owner [18]. In consequence, descriptions of knowledge-intensive business processes should include the full name of the specific expert[5] who is the source of a specific piece of knowledge in order to stress the fact that only that individual has access to that knowledge and must be contacted in order to receive the current version thereof.

---

[4]"BPs can then be considered as knowledge-in-action or actionable knowledge, thus reinforcing the need for better integration of KM and BPM, especially in the context of knowledge-intensive BPs" [46, p. 183].

[5]The term *expert* has a narrower meaning than the term *knowledge worker*. Experts are the best-educated and the most-experienced among the knowledge workers. The status of the knowledge workers stems not from being an expert, but rather, from the nature of the performed work, e.g. that of a lawyer, designer, consultant, medical doctor, analyst. In turn, the status of an expert is derived not just from one's expert knowledge, but also one's reputation in the field, metaknowledge, and wisdom. All knowledge workers make use of tacit knowledge tied to their work, but the vale of the knowledge of the expert is also determined by the way in which the expert obtained the knowledge in question, that is, his history of professional experience. Specific tasks are often assigned to experts who have worked on similar projects in the past.

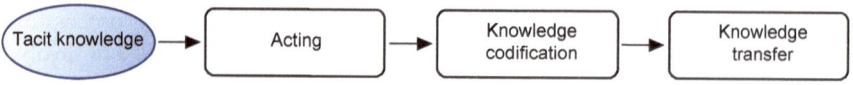

Fig. 3.9 The uncovering and transfer of tacit knowledge in traditional process management. *Source* Author's own elaboration

Traditional business process management assumes that processes are performed without deviations, in accordance with a standard process "as of today" accepted by the management of the organization. Only process owners and process leaders are authorized to introduce changes to standard processes on the basis of needs analysis and the course and results of subsequent process performances. Within this concept, process performers themselves are prevented from using their own knowledge with a view to raising the productivity and efficiency of process execution. As had been shown in Chap. 2, in the case of traditional business process management, the knowledge of process performers remains unused in the course of performance by design. It may be uncovered and used outside of performance in the Process optimization, definition, and modeling stages of the lifecycle (Fig. 2.1 in Chap. 2) or in the Process (re)Design stage (Figs. 2.4 and 2.5 in Chap. 2), if that is the decision of the process owners or process leaders. Because knowledge codification follows the use of knowledge in practice, the transfer of knowledge is possibly only upon the conclusion of process execution (Fig. 3.9).

Because tacit knowledge is the most easily revealed and transferred through taking action, this approach considerably limits the scope and lowers the efficiency of revealing knowledge[6] [5, 22]. As has been discussed in Sect. 1.1, the above method of uncovering tacit knowledge does not remove the negative results of the two fundamental problems with its discovery—that is, the unawareness of owning such tacit knowledge to begin with and difficulties in verbalizing it, as well as the reluctance to reveal and codify knowledge in isolation from action—but further deepens them instead.

In consequence of the principles underlining traditional process management, the creation of and access to new explicit knowledge lags behind due to organizational factors and decisions of the management. At the same time, traditional process management introduces the risk of using old, outdated knowledge—despite the organization experiencing the harmful effects of such use in specific contexts of performance.

---

[6]In the past, owners of chicken farms had to wait 5–6 weeks until the first feathers enabled them to distinguish between male roosters and female chicken. The egg farmers were interested in purchasing female chicks alone. They were intrigued by the news that some Japanese farmers were able to distinguish the sex of one-day chicks. Despite that fact that even poultry breeders are not able to distinguish between male and female sex organs in newly hatched chicks, the Japanese specialists were able to distinguish between the chicks at a mere glance. Hatcheries all over the world have sent their workers to training sessions held by Japanese specialists and then evaluated the level of proficiency of the newly trained employees. After months of theoretical and practical sessions the best American and Australian chick sexers were almost as good as their Japanese peers and were able to evaluate 800–1000 chicks with 99% accuracy.

By acknowledging tacit knowledge as a component which can be managed in the course of dynamic business process management, we are opening up new possibilities of the systemic verification and renewal, and, first and foremost, the revealing and codification of tacit knowledge. The tacit knowledge of practically all employees may be included in the systemic processes of knowledge management in the organization [48]. This requires organizations to empower employees to decide on the specific method of performing work (2nd principle of dynamic BPM) and document their work in the course of performance itself (3rd principle of dynamic BPM) in accordance with their best knowledge [22].

In effect, it is possible to manage the entire knowledge used in the course of process execution, even the tacit knowledge revealed in the course of operations. This primarily pertains to using and verifying, but also to codifying and transferring knowledge. In all of the abovementioned scopes, the analyzed performances of dynamically managed processes contain both components of explicit knowledge, as well as components of tacit knowledge. The scope and the level of detail of analyses of used or created knowledge depend on the level of detail of the documentation created by the process performers in the course of their work and on the scope of analyses of information on the context of performance. If IT systems in which work is documented will allow for the automatic entry of information from used social media portals, mobile devices, e-mail correspondence, or other forms of information exchange or making notes, as well as the entry of automatically created information from devices connected to the Internet of Things (IoT), it will be possible to recreate the full context of a given performance, including social relations [49]; ISIS Papyrus [50]. When the documented data will be systematized on an ongoing basis (e.g. based on the time of performance, the scope of cooperation with particular business partners, or their mutual ties), the acquired knowledge will be available for the needs of analysis even without the participation of the individuals who participated in its creation. In effect, it is possible to considerably broaden the scope of revealing and subsequently transferring tacit knowledge.

Within dynamic business process management, explicit knowledge is created in the course of process discovery and modeling. Subsequently, the entire body of tacit and explicit knowledge is used in the course of process execution. In accordance with the 3rd principle of dynamic BPM, in the course of performance all of the used knowledge is revealed. In other words, there is no justification for distinguishing between an Acting stage and a Knowledge creation and codification stage (Fig. 3.10). Knowledge created or discovered in this way may undergo further generalization, refinement, recording, and distribution with a view to its broad application in the processes performed in the organization. This pertains to both the entire body of "old" knowledge used in the course of process execution, as well as the "new" knowledge created or acquired in the process. This allows for:

- on-line analysis of the performed processes with the use of e.g. traditional control mechanisms, tools from the field of ML and AI, or expert systems [51];
- ex-post analysis of the performed processes with the use of e.g. process mining techniques or *Big Data* analysis techniques [52].

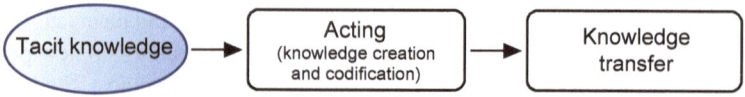

Fig. 3.10 The uncovering and transfer of tacit knowledge in dynamic process management. *Source* Author's own elaboration

Knowledge management possibilities offered by dynamic BPM have been used by the author in the course of performing an implementation of knowledge management with respect to knowledge used and acquired in the course of performing processes in a hospital ER unit. In the course of analyzing ER procedures and specific treatment processes documented in a Hospital Information System (HIS), the author has analyzed the clinical pathways which were the most common in the treatment of selected diseases. Thanks to the codification of the diagnosed diseases with the help of the common ICD10 classification, it was possible to make a thorough comparison between medical procedures codified in accordance with the ICD9 classification performed by different medical doctors who in the course of diagnosis suspected the same disease. It became possible to compare the approaches in terms of their outcomes, that is, the improvement of the patient's condition, duration, costs, resources, etc. One of the results of the analysis was the discovery that in 90% of cases when the medical doctors suspected a specific group of conditions in a specific group of patients, following an ultrasound scan it was crucial to perform a more thorough examination - a CT scan - within the next 24 h. The change of the standard clinical pathway introduced in result of the analysis have allowed the medical doctors to eliminate unnecessary waste of time, which threatened patient health, as well as reduce unnecessary costs, which de facto resulted from performing needless ultrasound scans. Knowledge which has been unavailable to date on the specific scenarios of treatment used by a broad spectrum of medical doctors has now become explicit, open to evaluation, and available for quick dissemination. [53]

In the case of traditional, static process management, the concept of knowledge management is often detached from the daily operations of the organization and reduced to the refinement of explicit knowledge. One solution which enables the systemic and natural uncovering and use of tacit knowledge is the combination of dynamic business process management with knowledge management. It allows for the creation of competitive advantage in the organization on the basis of the ongoing, constant uncovering of knowledge on process execution and the most rapid use of the said knowledge with a view to process improvement [54].

In 2001, in the course of a study on experiences with knowledge management, Piotr Płoszajski's team conducted an analysis of companies which effectively use the knowledge of their employees, clients, and suppliers. The analysis distinguished between several fundamental factors for success. The first of these factors was including learning in the business processes of the organization (just-in-time learning) [55]. The use, verification, and creation of knowledge must be situated in the hands of the process performers themselves [13]. It is not a coincidence that it is in Jenning's and Haughton's book on speed as the decisive asset of companies of the future that we find the statement that decisions to be fast and accurate, should be taken as close to the place of real action, as it is possible [56]. To ensure efficient knowledge management, a methodological approach encompassing both process management and knowledge management provides a much larger chance of success than implementing projects from scratch which are limited to knowledge management alone [57, p. 7].[7] However, it requires an approach to process management understood as dynamic business process management.

As has been demonstrated in Chap. 2, in traditional process management the knowledge of process managers is de facto left unused—wasted in a systemic manner. For this reason, in order to reap the benefits of the synergy between knowledge management and process management, it is essential to enable the use of knowledge not just in the modeling stage and during the ex-post analysis of process execution, but first and foremost in the course of performance itself. According to Remus [58], it is crucial to resort to process-oriented knowledge management and at the same time consciously shape business processes with a view to their use and support of knowledge management. Heisig [59] points to the possibility of including the demand for knowledge as early as in the process modeling stage, as well as including in knowledge-intensive business processes the evaluation of knowledge on the basis of changes to and deviations from the standard process, which surfaced in the course of performance. According to Gronau [60, p. 45], in order to achieve synergy between process management and knowledge management, knowledge-intensive business processes should be performed in an environment accounting for the fact that:

- value created in the course of a knowledge-intensive process can only be created thanks to the use of knowledge by direct process performers, but at the same time, process execution itself creates additional value for knowledge management;
- value crated in the course of a process depends on the creative and innovative use of knowledge [37];
- individuals who participate in process execution contribute their diverse knowledge, including their individual experience [59];
- the time to create or acquire knowledge in the course of performing a business process is usually very short. At the same time, the duration for which knowledge remains up to date is also limited. For this reason, it is essential to quickly transfer and use the created knowledge before it becomes obsolete (or before it falls in the hands of the competition);
- process performers must have a systemic—organizational—broad scope of autonomy with respect to acting and making decisions;

---

[7]"It could be seen that a combined methodological approach for BPM and KM in a company is much more promising than starting from scratch with a pure KM project" [57, p. 7].

- the course of a knowledge-intensive process cannot be strictly defined nor considerably limited before performance;
- apart from performance indicators measuring the productivity or operational efficiency of knowledge-intensive processes, there must also be defined other indicators on the use, verification, and creation of knowledge.

The above requirements are 100% applicable to an organization in which processes are managed in accordance with dynamic BPM. The convergence of the models of knowledge management and dynamic business process management is not coincidental (Fig. 3.11). It is the logical consequence of allowing, within dynamic business process management, for the genuine shaping of the performed processes not just by the organization's management, but by all of its knowledge workers—not through one-off actions (as the result of audits, the end of the financial year, or changes

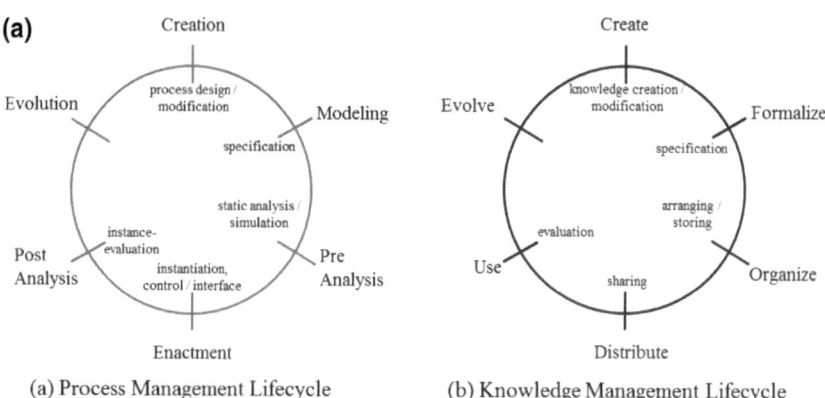

(a) Process Management Lifecycle   (b) Knowledge Management Lifecycle

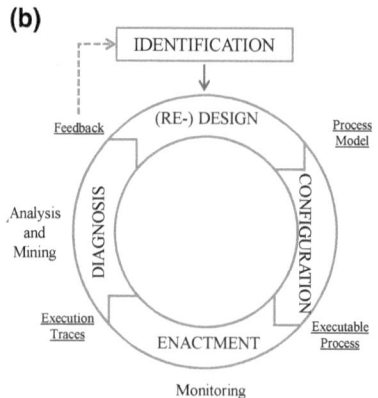
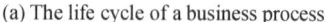
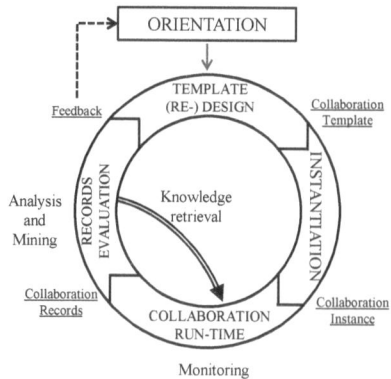

(a) The life cycle of a business process   (b) Collaborative knowledge work life cycle

Fig. 3.11 The convergence of the models of knowledge management and process management. *Source* **a** Jung et al. [61, p. 23]. **b** Di Ciccio et al. [62, p. 2, 5]

in ownership), but in the course of systematic, constant process improvement and the creation and verification of knowledge.

3.3.2 Proposal of Updating the Process-Oriented Model of Knowledge Management

Davenport's and Prusak's [5] process-oriented knowledge management model presented in Fig. 3.7. clearly distinguishes between the Knowledge acquisition and Knowledge codification processes. Furthermore, it does not include a Knowledge use process with a view to decision making, nor a Knowledge use process with a respect to process execution. For this reason, multiple authors have postulated its supplementation with a Knowledge use process, which would encompass knowledge acquisition and use, as well as the transformation of knowledge into business decisions [30]. At the same time, Buckley and Carter [63] proposed to simplify the Knowledge management process by distinguishing between only two processes within it: the Knowledge transfer process and the Knowledge creation process. However, the above proposals to improve the process-oriented knowledge management model do not solve its fundamental practical problems:

- disentangling implementations of knowledge management from the day-to-day operations of the organization (the performance of the fundamental process);
- overstating the role of technology and IT tools and underestimating the role of the knowledge workers.

The solution to such problems lies in implementing dynamic business process management. This approach ensures the possibility of the systemic harmonization of process management and knowledge management with the day-to-day operations of the enterprise by indicating process execution as a constant space for verifying old knowledge and a source of new knowledge, as well as a space for verifying the requirements for new knowledge. However, this requires us to introduce significant changes to the process-oriented knowledge management model with respect to the stages in which knowledge is acquired (created) and codified (Fig. 3.12). These

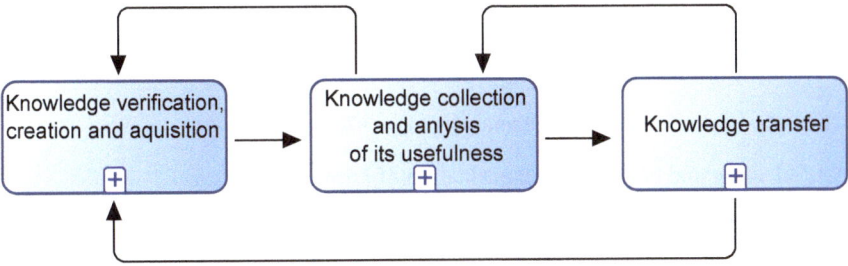

Fig. 3.12 The process-oriented knowledge management model. *Source* Author's own elaboration, on the basis of Davenport and Prusak [5], Sopińska and Wachowiak [30]

changes are introduced with a view to benefiting from the fact that in the course of process execution itself, dynamic BPM allows for:

- revealing tacit knowledge;
- creating new knowledge;
- acquiring knowledge from outside the organization;
- codifying knowledge.

It also enables us to redefine the role ICT solutions as tools subordinated to the dynamism and innovativeness of the knowledge workers, used with a view to:

- supporting work and revealing knowledge;
- publicizing and transferring knowledge;
- evaluating the results of using knowledge on the basis of objective, automatically generated performance indicators.

The proposed scopes of operation within the framework of each stage, or rather, specific processes comprising knowledge management, are presented in Table 3.3.

Actions which were not possible within the boundaries of traditional business process management are highlighted in italic.

- Knowledge verification, acquisition, and creation

    - Knowledge verification in the course of process execution
      in the case of dynamic BPM, the results of each process execution may be compared with the standard process not just with the use of performance indicators (time, cost, resources used, etc.), but also by way of evaluating the performance in the specific business context and in cognizance of the personal experience of specific process performers.
    - Knowledge disclosure
      in accordance with the 3rd principle of dynamic BPM, the acts of disclosure and codification encompass not just information on process execution, but also information on its full business context and the experience of the process performer or the entire team of performers. Within dynamic business process management, knowledge codification is performed in a systemic manner in the normal course of performance itself. In effect, it does not require the existence of additional procedures and processes and does not impose additional costs or workload on the employees or the management. In consequence, it allows for considerably broadening the scope of knowledge management in the organization with a view to encompassing not just selected experts, but practically the entire personnel of the organization.

- Knowledge collection and the analysis of its usefulness

    - Selection and contextual subordination of knowledge
      as has been noted in Chap. 2, modern transactional and process management IT systems, such as MRPII, ERP, CRM, HIS, EMR, BPMS, or CMS, but also information exchange systems (such as customer communications management—CCM), enable the acquisition of information from multiple channels

Table 3.3 Scopes of operation within the framework of processes comprising knowledge management

The process-oriented knowledge management model	Knowledge verification, acquisition, and creation	*Knowledge collection and the analysis of its usefulness*	Knowledge transfer
Scope of operations	Use and *verification* of old knowledge	Verification of the validity of knowledge codification (both old and new)	Distribution and acquisition of information and knowledge
	Search for and the acquisition of information and knowledge	*Selection and contextual subordination of information*	*Automatic distribution and acquisition of information and knowledge*
	Creation and *verification* of new knowledge	Analysis and evaluation of the usefulness of old knowledge	Knowledge transfer in the course of formal operation
	Disclosure (codification) of the created new technology	*Preparation of knowledge for distribution*	Knowledge transfer in the course of informal contacts between knowledge workers
Dynamic BPM	Process performance	Analysis of ongoing and completed processes	Process discovery and management (management of process descriptions and process expert)

Source Author's own elaboration

of information flow, including social media applications, messaging software, and e-mail clients. Following the rejection of information noise, the remaining information is organized and tied to data from fundamental transactional and process management systems. Knowledge which had been tacit to date may become a very crucial source on e.g. informal sources of information, factors influencing the decision-making process, or the actual course (sequence) of events.

– Preparation of knowledge for distribution

  because the form of presentation of knowledge has considerable influence on the efficiency and pace of its absorption, knowledge should be prepared for transfer to specific groups of recipients as early as in the knowledge collection stage.

• Knowledge transfer

– Automatic distribution and acquisition of information and knowledge
  as has been noted in Chap. 2, one of the problems with respect to knowledge management is "information overflow," and another—the unawareness of lacking
  in knowledge [28]. In accordance with subchapter 1, another problem pertains
  to the codification of knowledge, which in traditional process management is
  perceived as an additional, unnecessary duty to be performed upon completion
  of the work itself. In dynamic business process management, in accordance with
  the 3rd principle of dynamic BPM, work documentation is performed as an integral component of business process execution. In effect, it offers the possibility
  of the ongoing transfer of new, practical knowledge to experts and knowledge
  workers.

Table 3.3 includes a proposal to harmonize knowledge management with dynamic
BPM. The scope and sequence of actions within specific processes comprising knowledge management (KM) and business process management (BPM) correspond with
each other to a significant degree. Because both knowledge management and business process management are on principle ongoing processes, it should not matter
whether a specific organization begins this harmonized process from the traditional
Knowledge transfer process (KM), which corresponds to Process identification in
BPM, or whether it begins with process discovery in the Knowledge verification,
acquisition, and creation stage (KM) on the basis of data from IT systems, which
corresponds to Process execution in BPM.

3.3.3   Relations Between Nonaka's and Takeuchi's Knowledge Management Model and Dynamic Business Process Management

The abovementioned analogies are also present between Nonaka's and Takeuchi's
knowledge management model and dynamic business process management
(Table 3.4).

The Dissemination of tacit knowledge stage of dynamic business process management should account for the possibility of process performers uncovering tacit knowledge in the course of performance itself. Thanks to the documentation of subsequent
performances in accordance with the 3rd principle of dynamic BPM, such knowledge becomes explicit, that is, possible to be analyzed and distributed throughout the
organization. At the same time, changes to the performed processes document the
dynamic search for innovative ideas (in the Creating concepts phase), and the course
and the results of processes contain an objective evaluation of changes introduced
in this way (in the Justifying concepts phase). They then form the basis of updating standard models and process descriptions (in the Building an archetype phase)
in dynamic BPM, which is limited to the management of the organization (process

Table 3.4 Comparison between Nonaka's and Takeuchi's knowledge management model and knowledge management under the conditions of dynamic business process management

Phase	Knowledge management by Nonaka and Takeuchi	Knowledge management within dynamic BPM
Phase I	Sharing tacit knowledge	1. Process identification—knowledge on the standard operations of the organization 2. *Documentation of the actual performance of work—knowledge on limited experiments and their results*
Phase II	Creating concepts	1. Planned operations by the organization's management 2. *Changes introduced in the course of process execution itself (systemic, active, limited experimentation)*
Phase III	Justifying concepts	1. Analyses performed according to the schedule of the organization's management 2. *Analysis of the results of performances of individualized processes*
Phase IV	Building an archetype	Changes to the standard process introduced by process owners
Phase V	Cross-leveling knowledge	1. Standard innovation processes in the organization 2. *Communities of practitioners based on knowledge bases derived from process executions* 3. *Databases of best and wrong practices*

Source Author's own elaboration, on the basis of Nonaka and Takeuchi [3]
Italic marks possibilities offered by dynamic BPM (and its supporting IT solutions), which were not available in traditional business process management

leaders and owners). The dissemination of knowledge (in the Cross-leveling knowledge phase) may be attempted not just by way of formal (static) operations of the organization, but also informally (dynamically): by social media applications, messaging software, consultations with experts, etc., actions which can be collectively termed *social* BPM.

Table 3.4 points to the two-track nature of both knowledge management (KM), as well as dynamic BPM. "Static" steps, which fall in line with traditional business process management, are planned in a top-down manner and are managed in a way which has no direct connection to performance itself. In turn, "dynamic" steps are performed by knowledge workers in the context of specific business process executions. This two-track nature is clearly visible in the analysis of relations between Nonaka's and Takeuchi's knowledge management model and the model offered by dynamic business process management, which is presented in Table 3.5.

Table 3.5 Relations between two knowledge management models—Nonaka's and Takeuchi's knowledge management model and knowledge management within dynamic business process management

The process-oriented knowledge management	Knowledge verification, acquisition, and creation	*Knowledge collection and the analysis of its usefulness*	Knowledge transfer
Nonaka's and Takeuchi's model	*Sharing tacit knowledge (dynamic)*		Sharing tacit knowledge (static)
	Creation concepts (dynamic)	Creation concepts (dynamic)	
	Justifying concepts (dynamic)	Justifying concepts (dynamic)	
		Building an archetype	Cross-leveling knowledge (static)
	Cross-leveling knowledge (dynamic)	*Discovering concepts used in process performance (dynamic)*	
Dynamic BPM	Process performance	Analysis of ongoing and completed processes	Process discovery and management (management of process descriptions and process experts)

Source Author's own elaboration, on the basis of Nonaka and Takeuchi [3]
Italic marks possibilities offered by dynamic BPM (and its supporting IT solutions), which were not available in traditional business process management

One should note the possibilities of the dynamic uncovering (codification), creation, and dissemination of knowledge in the course of performance itself, which stem from the principles of the concept of dynamic BPM. Within this concept, the role of the employees is not limited to the efficient performance of business processes, but also encompasses the creation and use of knowledge during performance itself, with a view to providing value for the client and the organization. In this way, the intellectual capital of knowledge workers can not only begin to work for the benefit of the organization (in accordance with the 2nd principle of dynamic BPM), but the knowledge derived therefrom is immediately revealed and ready to be transferred and used multiple times in the organization (in accordance with the 3rd principle of dynamic BPM). Another requirement of dynamic BPM rests in the principle that actions with a view to revealing tacit knowledge, which encompass its creation, use, and verification, be performed routinely in the course of the organization's fundamental operations. As noted before, the dynamic creation and revealing of knowledge should function alongside process management in the scope known from traditional business process management. In the model presented in Table 3.4, both tracks of process development and improvement function in parallel, creating conditions for efficient knowledge management and allowing organizations to share the results in the statically performed Building an archetype stage and the statically and dynamically performed Cross-leveling knowledge stage

3.4 Process Execution as a Source of Knowledge

Process execution in accordance with the concept of dynamic BPM enables process performers to tailor the course of a given process to the requirements of a specific situation. With the use of ICT solutions, it is possible to combine dynamic business process management with knowledge management with a view to creating a harmonious organizational and IT solution which draws on knowledge derived from process execution. As shown in Fig. 3.13, this requires organizations to incorporate knowledge management in the performance of business processes [38].

Process improvement is a response to the requirements of specific process executions, the pressure exerted in a specific situation, the requirements of a specific client, etc. In this case, it is irrelevant whether such improvements are the result of knowledge derived from outside the organization or the ideas and intuitions of the personnel ("necessity is the mother of invention")—be it a single individual or an entire team. What is important is that thanks to the 2nd and 3rd principles of dynamic BPM, the organization is able to considerably broaden its capacity to create new knowledge and discard old knowledge, as well a take responsibility for this process. It should be noted that this is a source of practical knowledge, which is derived from the day-to-day work of the intellectual capital of the organization—not the

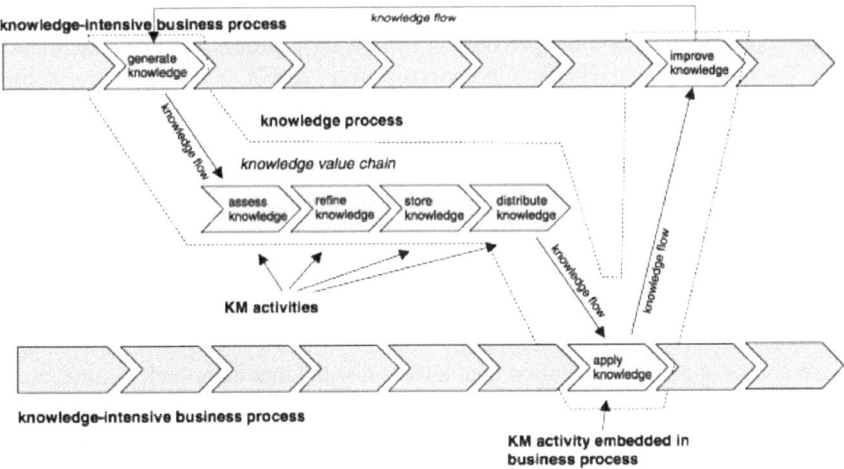

Fig. 3.13 Relations between process execution and knowledge management processes. *Source* Author's own elaboration, on the basis of Remus and Schub [38]

result of theoretical knowledge purchased from consulting companies, which, valid as it may be, may still prove impossible to implement in a specific organization; nor even the result of knowledge created in the course of one-off actions with the aim of updating process standards. We are speaking of practical knowledge, which is often the result of unforeseeable changes to customer requirements, the market, and technology, which in dynamic BPM is acquired from an ongoing source: with the use of the entire (or the majority) intellectual capital of the organization. This ongoing source allows us to:

- acquire up-to-date knowledge, which is implementable in day-to-day operations [64];
- verify and improve knowledge on an ongoing basis [65].

If not for dynamic business process management, the entire concept of knowledge management would often become disentangled from the day-to-day operations of the organization, which absolutely has no case under dynamic BPM. At the same time, if not for the implementation of process-oriented knowledge management, the implementation of dynamic business process management is a futile undertaking. Its benefits would be thwarted regardless, as knowledge created with the help of dynamic BPM would not be identified and shared with a view to its further use. The aim of the harmonious combination of knowledge management with process management is the ongoing, constant acquisition of knowledge on process execution and its instant use with a view to process improvement.

As has been discussed in Sect. 3.1, this requires the introduction of changes to the standard process lifecycle proposed in the Process Mining Manifesto [47, pp. 6–7]. The new, generalized process lifecycle allows not only for "improving" performed processes by was of limited experiments (small improvements), but also for the ongoing analysis and communication of such improvements throughout the organization—for employees and teams participating in the modeling, improvement, implementation, and ongoing performance of business processes. The combination of ICT solutions from the fields of process discovery and *Big Data*, as well as source transactional systems (ERP, CRM, MRP II, HIS, EMR), social media portals, the Internet of Things, and telecommunication software (e.g. e-mail and online messaging applications) enables the identification and use of the endless stream of knowledge with a view to ongoing process improvement and foreseeing the direction of future changes [66]. In practice, the automatic logging of such data in the event logs of IT systems supporting process execution on the 4th or at the minimum 3rd maturity for event logs enables us to fully recreate the performance of a given process [67, 47]. In effect, by assessing multiple process executions, it is possible to determine:

- the course of process execution in connection with other data, such as time, costs, or key performance indicators for the results of the process or its components;
- contextual scenarios for process execution, e.g. a sales process dependent on the type of goods, the character of the client, the time of day; or investment processes dependent on size and location (e.g. housing, infrastructure, office space) [68];
- deviations from the standard process and their effects (exceedingly positive or negative results);

- process patterns for the performance of fundamental sub-processes (e.g. decision-making sub-processes), which may be used in multiple processes;
- experts for selected scopes of the process (e.g. financial, production, logistics) and selected scopes of process management (e.g. optimization of efficiency or implementation of innovations) [69];
- new factors or criteria affecting the efficiency and scope of processes, which may need further codification and analysis of their influence (e.g. the influence of engagement in social media or the practical use of the Internet of Things);
- benchmarks (ongoing and prognostic) which exceed simple statistics.

At the same time, it is possible to acquire valuable knowledge in the specific field by including as early as in the process modeling stage:

- data objects containing knowledge valuable to the process performer and available in the course of performance itself, e.g. tips on a particular subject, checklists, process patterns, best practices, the option to consult with communities of practice (the equivalent of "phone-a-friend" or another such lifeline);
- attributes and data objects representing knowledge acquired in the course of performance, e.g. selections from checklists, information on experts, tasks performed outside of the standard process and their fundamental parameters.

The capability to oversee the course of performance in combination with other data, such as time, costs, or performance indicators for the results of the entire process or its parts, allows us to perform analyses with a view to process optimization. At the same time, we are able to acquire knowledge on the broader context of the actual performance of processes. In order to use such knowledge with a view to raising efficiency, minimizing risk, or ensuring competitive advantage, we first need to subject knowledge to standard knowledge management processes.

To conclude, the implementation of dynamic BPM may become an ongoing, virtually free source of knowledge—even more essential knowledge than explicit knowledge or tacit knowledge or acquired from outside the organization. By drawing on this fact, one can, on the basis of the works of Kim et al. [37], propose a knowledge creation architecture based on business process execution.[8] The diagram presented in Fig. 3.14 includes the following knowledge flows:

- knowledge absorption (KA)
- knowledge extraction (KE)
- knowledge implementation (KI)
- knowledge deployment (KD)

In this model, knowledge is derived in the course of process execution in a twofold manner:

- explicit knowledge—in a formal manner, on the basis of information on process execution containing process results (e.g. predefined process indicators) and the results of using knowledge in the course of performance (e.g. predefined knowledge management indicators);

---

[8]Here, of course, business processes should always be understood as dynamic business processes.

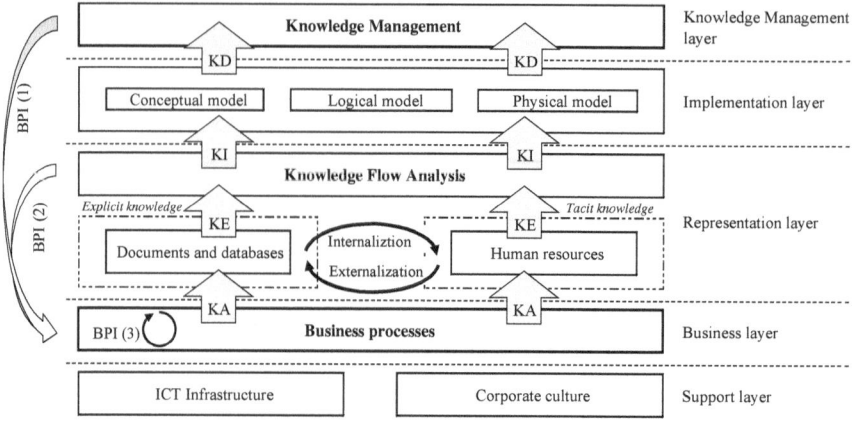

Fig. 3.14 General framework of knowledge management in dynamic business process management. *Source* Author's own elaboration, on the basis of Kim et al. [37, pp. 260–276]

• tacit knowledge—in an informal manner, on the basis of individual and collective experience.

In the knowledge management model in dynamic business process management, there are three independent organizational and substantive levels of knowledge oriented business-process improvement (BPI):

1. BPI (1)—improvements resulting from actions by the management of the organization, performed in the form of:

   – one-off actions, resulting from e.g. changes in ownership or changes in the organizational strategy;
   – periodic overviews (audits) performed in accordance with the implemented quality management methodology, as well as internal and external regulations;

2. BPI (2)—improvements resulting from the ongoing analysis of the results and pathways of process executions an the efficiency of the flow and the use of knowledge, performed by process owners and managers within the limits of their privileges. In justified cases the said individuals may initiate improvements on the BPI (1) level;

3. BPI (3)—improvements performed in the form of improvement attempts (limited experiments) by process performers in accordance with the 2nd principle of dynamic business process management. Their analysis on the level of knowledge flow analysis may become the basis of initiating process improvements on levels BPI (1) or BPI (2).

The use of the possibilities offered by knowledge management under the conditions of dynamic business process management leads to a situation in which, on the one hand, work is considered completed when the practical knowledge used or created in its course is revealed, and—on the other—the evaluation of the work in question is based on e.g. indicators measuring the efficiency of using available knowledge and creating new knowledge. In effect, we are dealing with a fully institutionalized process of measuring knowledge distribution, which is the foundation of knowledge management in the organization. Objective key performance indicators (KIP) based on measurements of specific processes prevent mistakes resulting from rewarding ideas which are subjectively the most innovative, albeit which are impossible to implement or destructive in nature. This allows organizations to eliminate discrepancies between sharing knowledge and internal competition. In the case of dynamic BPM this problem is nonexistent, as internal competition is rooted in actual effects of work, which are verified in the most objective manner possible: by the client.

3.5 Execution of the Concept of a Learning Organization in Dynamic Business Process Management

3.5.1 The Concept of a Learning Organization

The concept of a learning organization first surfaced in the 1990s. Among the various definitions of a learning organization this book will make use of just two. One of the most popular definitions of a learning organization was formulated by Peter Senge: "Learning organizations [are] organizations where people continually expand their capacity to create the results they truly desire, where new and expansive patterns of thinking are nurtured, where collective aspiration is set free, and where people are continually learning how to learn together" [70]. The second definition was formulated by Sikorski and reads as follows: "A learning organization is a maximally flexible organization, in which routine, habits, and stereotypes do not replace the dynamic reality" [71, pp. 29–35]. To generalize: a learning organization is an optimally flexible organization, in which routines, habits, and stereotypes change under the influence of the knowledge of the dynamic reality and the perceivable future. According to Lassey, the key to understanding a learning organization is unending development. Assuming that the learning process is a modification of behaviors, a learning organization must be capable of modifying its own patterns of behavior [72]. In effect, it must be able to adapt, transform, and to develop itself [71, p.30]. Then it will have perfectly implemented processes of organizational learning, which work on an ongoing basis. This is a good point of departure for looking at an organization from a somewhat different perspective: that of organizational learning (Fig. 3.15).

In Senge's model [70, pp. 6–10], building a learning organization is predicated upon the five following disciplines:

- systemic thinking;
- personal excellence;
- thought models;
- shared vision;
- team learning.

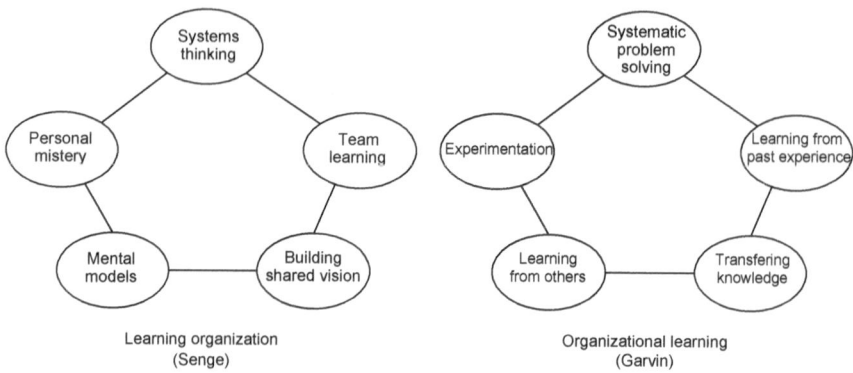

Fig. 3.15 Two approaches: a learning organization (Senge) contrasted with organizational learning (Garvin). *Source* Author's own elaboration, on the basis of Jashapara [4]

According to Garvin, a learning organization should be proficient in generating, acquiring, and sharing knowledge, as well as implementing the newly acquired knowledge in ongoing operations [4, pp. 180–183].

Learning in itself does not translate to building competitive advantage or generating better-than-average results. Both are achieved only after imbuing the learning process with a proper focus. Organizational learning may be adaptive (single-loop learning) or generative (double-loop learning) in structure [73, 70]. Single-loop learning translates to "doing the same thing better," whereas double-loop learning—"doing the same thing differently or doing another thing better" [4]. In the hypercompetitive environment, organizations are forced to learn in a generative (double-loop) manner, as only then are they capable of keeping up with unforeseeable qualitative changes in customer requirements and the competition, without neglecting "doing the same thing differently." (Fig. 3.16).

From the perspective of Garvin's model, we can distinguish between two fundamental methods of learning on the level of the individual, the team, and the entire organization:

- shaping—learning through experience and trial and error, or, in other words, active experimentation in solving ongoing problems and daily challenges;
- modeling—drawing on the experience of others, or education and the observation of other teams or organizations and adopting their methods of operation.

Tsang was on point when he stated in this context that organizations learning from practice will automatically gravitate towards making improvements in their performance, as long as the process is accompanied by appropriate knowledge [70, p. 39, 75]. His thesis has been confirmed by research held by Jashapara [4, p. 196]. At present, this method of learning is increasingly singled out as the most efficient. Nevertheless, it remains necessary to solve the problem of gaining, analyzing, and circulating experiences gained from active experimentation and knowledge acquisition, including the knowledge obtained by observing other organizations [76]. The

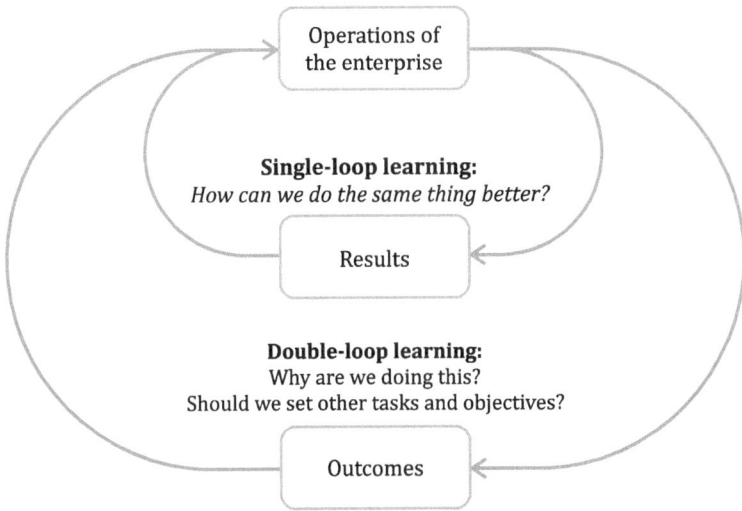

Fig. 3.16 Single-loop and double-loop learning. *Source* Author's own elaboration, on the basis of Gladstone [74]

sought-after organizational system should within processes pertaining to knowledge management share several features in common with the general knowledge lifecycle model:

1. The creation of new knowledge
   Employees should be able to make individual choices on how best to approach their work. Organizational procedures (e.g. ISO quality management systems) and process models should enable employees to search for the most efficient solutions, or de facto allow for active experimentation, which is present in Garvin's model (1993).
2. The analysis of created knowledge
   The management should be able to monitor changes introduced to work performance on an ongoing basis, as well as to measure the results of work in an objective and quantifiable manner, both on the level of comprehensive customer support (individual orders, contracts, or products), as well as on the level of particular tasks. Only then will it be possible to identify those experiences which should be shared throughout the organization (best practices), as well as to identify those behaviors which should be avoided (wrong practices).
3. The dissemination of knowledge
   The process of organizational learning should not be limited to collecting knowledge and information, but should also allow for their rapid dissemination throughout the organization with the aim of using knowledge in business practice with a view to gaining competitive advantage.

A system with the above features allows us to shape and model all of the operations within the organization. Knowledge on customer requirements and the efficiency of particular adaptation mechanisms should be stored in shared databases and verified on a continuous basis, in the form of e.g. documentation of performed processes or patterns to be adapted and modified for actual use. Such well-applied knowledge quite frequently guarantees competitive advantage in terms of reaction time and the proper implementation of processes which adapt to changes outside of the organization. In an ideal situation, an organization should possess knowledge which allows, with great probability, to anticipate, or at the very least, closely follow the changes that are happening or are about to happen outside the organization [71, p. 66].

3.5.2   Practical Implementation of the Concept of a Learning Organization in Dynamic Business Process Management

Below is an overview of the process of knowledge management in a learning organization managed with the use of dynamic BPM. Each newly-hired employee receives fundamental data and information on the organization and its specific character. Usually, apart from introductory training sessions, such information is provided in the course of a senior employee/novice mentor relation. Upon commencement of work, the junior employee begins to create individual knowledge. Should the employee leave the company, they also irrevocably takes away their individual knowledge, which to a large degree has the form of tacit knowledge that neither the owner nor the managers are able to absorb and keep in the organization [77].

The situation is different in organizations managed in accordance with the principles of dynamic BPM. The process owner is responsible for planning/designing the process, as well as for training the process performers. In other words, the process owner is the one who shares knowledge with the novices. Thanks to the possibilities offered by dynamic business process management, after the preliminary period of familiarizing new employees with the courses of business processes which comprise the organization's collective knowledge, the (knowledge) workers then contribute to the creation of such collective knowledge on an ongoing basis through the identification and selection of new solutions, as well as the verification of existing knowledge in the course of actual day-to-day operations. Such ongoing verification of existing knowledge (in accordance with the 2nd principle of dynamic BPM) is fundamental. Without it, in the age of rapid technological changes, as well as changes to the organization's environment, it could easily turn out that the organization is using old and outdated knowledge. In consequence, the ability to create and verify knowledge on an ongoing basis is a fundamental skill, which allows companies to maintain their capability to both change and react to changes. We are not speaking of a one-off action, of restructuring, reengineering, or similar provisional measures, which are usually unrelated to the generation of added value for the customers and aimed at restoring the ability to fulfill customer needs. Instead, what we have in mind are continued actions pertaining to the fundamental operations of the organization, which enable it to adapt to changing conditions. Such conditions include changing customer requirements, the owners, and the personnel alike (indeed, meeting the expectations of one's personnel may be as crucial as fulfilling the demands of the customers from the point of view of motivating good performance).

Within dynamic BPM, the ability to change and to generate change is constant and inscribed in the organization's ongoing operations. It fulfills all of the requirements put forward by Drucker or Hammer with respect to the "institutional ability to change." By way of verifying organizational knowledge on an ongoing basis and attempting to introduce innovations which would increase efficiency and provide the company with competitive advantage, dynamic BPM creates and institutionalizes the organization's potential to self-reform. The key to success lies not in being able to predict the future, but rather, the continuous adaptation of the principles of operation, with a view to facing an unforeseeable and surprising future [78, p. 1]. By introducing principles which allow for the dynamic modification of processes, organizations inextricably combine their fundamental operations with their day-to-day capability to introduce innovations, generate knowledge, and change. Because process performers are able to change processes dynamically, the entire system of business management opens itself up to the creative initiatives of employees without introducing the risk of chaos associated with the uncontrolled change of the principles of operation. Furthermore, with the capability to monitor the effects of changes, we can enrich the collective knowledge of the organization with practices and solutions which provide the best results.

Now we can indeed see Hammer's vision of what it means to be a process-oriented organization; one in which process improvement is neither secondary nor peripheral, but central to the task of management itself. This is what Hammer called the "deep system of management," which monitors, administers, adjusts, and reforms the surface system with a view to generating value for the customer [27, p. 162]. However, it is not a separate, external system which, apart from generating additional costs, may easily begin to be perceived within the organization as another bureaucratic duty impending normal work. Instead, it has the role of enabling genuine day-to-day improvements and adaptations introduced in the course of analyzing process execution. The organization's body of knowledge on the best practices which are currently in operation, as well as on the direction and methods of their modification, comprises the organization's property. At the same time, tacit knowledge is being minimized. IT systems and their databases, which are responsible for dynamic business process management, make virtually all of the collective knowledge of the organization accessible to all employees. It goes without saying that in such a situation, even when key employees leave the organization, practically all of their "personal knowledge" remains in the organization and remains its property by default, regardless of whether the organization is managed traditionally or operates as a virtual network. There is just one condition: the management board, or the "instigator" of a network company, should consequently enforce the use of dynamic BPM tools and the principles of documenting all operations.

An organization managed in accordance with the principles of dynamic business process management by definition fulfills the requirements set before a learning organization. All of the employees of such an organization—or at least a wide range thereof—create collective explicit knowledge on an ongoing basis.

3.6 Conclusions: The Integration of Process Management with Knowledge Management

Within standard knowledge management processes, in the course of process execution knowledge is first identified and evaluated, and then, generalize, refined, recorded, and distributed—with a view to its broad application in the processes performed in the organization. This pertains to both knowledge used in the course of process execution, as well as to new knowledge created or acquired therein. It allows us to close out the knowledge lifecycle in the organization with the most objective possible evaluation of its usefulness—namely, the evaluation of the client, the target recipient of the results of process execution. Organizations must learn how to make use of this evaluation with a view to deciding on the methods of adapting to changing conditions of operation. Failing that, their fundamental competences may turn into fundamental limitations [4]. If not for dynamic BPM, the entire concept of knowledge management would often become disentangled from the day-to-day operations of the organization, as forcing process performers to passively use existing knowledge rids the organization of the capability to verify knowledge, as well as to search for and validate ideas on an ongoing basis. At the same time, if not for the implementation of process-oriented knowledge management, the implementation of dynamic business process management or the use of process mining tools is a futile undertaking, the benefits of which would be thwarted regardless.

This chapter answers the question of *Is it possible to integrate (dynamic) process management with knowledge management, including the management of tacit knowledge?* in the affirmative. When implemented, dynamic BPM may serve as an ongoing, virtually free source of knowledge which thus far had been tacit to date—perhaps even more essential than explicit knowledge acquired through mechanisms of knowledge standardization and codification or acquired from outside the organization. The combination of dynamic BPM and knowledge management creates the necessary conditions for creating knowledge through unlocking the dynamism ("they want") and the creativity ("they are searching for") of a wide range of process performers, as well as enabling the management of all the knowledge at hand, ant not just knowledge which is explicit in nature.

Incorporating the management of the entire knowledge of the organization into process management is fundamental factor behind success in the organization, providing it with:

- the constant, institutional readiness to change thanks to the day-to-day search for new solutions undertaken by a wide range of employees;
- the capability to constantly reveal, create, and verify knowledge within the organization on the basis of the experience of a wide range of employees;
- the capability to use and share tacit knowledge, which thus far had been available only to its direct owner;
- the capability to use the created knowledge on an ongoing base and at a faster pace than the competition thanks to the systematic use of collective knowledge.

Because in dynamic BPM work is considered completed only after having been documented, virtually all of the knowledge created in the organization is explicit in nature and available for rapid distribution throughout the organization. The concept of dynamic BPM, which is being developed since 2004, is not the first attempt at overcoming the limitations of traditional, static business process management and accommodating management to the requirement of the increasingly more hypercompetitive business environment [19]. The experience thus far in implementing dynamic BPM gives us hope that by genuinely using the dynamism of a wide range of employees, this solution will allow us to combine the efficiency of process management with the flexibility and openness to change of knowledge management. Organizations in which knowledge and processes are managed separately are fast becoming obsolete and cannot compete with those which empower their personnel to combine knowledge management with process management [79]. The generalization of traditional, static business process management to dynamic business process management allows us to harmonize and integrate the management of different practical methodologies and IT tools within a single cohesive concept. The goal set before organizations is the ongoing, constant acquisition of knowledge on process execution and its rapid use in process improvement on an organizational and extra-organizational level. This is possible only by the harmonious combination of knowledge management and (dynamic) business process management. Only then is it possible to acquire an verify knowledge deriving from all potential sources, including tacit knowledge accessed in the course of creative problem-solving and experimentation, following which knowledge is used and shared.

In conclusion, the question posed in the beginning of this chapter should be answered in the affirmative. Furthermore, it is independent of the adapted "process-based" or "case-based" methodologies and management tools. Arriving at a cohesive concept encompassing process management in its entirety and integrating it with knowledge management will be achieved (or rather: is being achieved at present) by multiple concurrent paths and from different points of departure.

References

1. Marshall A (1920) Principles of economics, 8th edn. Macmillan and Co., Ltd., London. First published: 1890. Retrieved from http://www.econlib.org/library/Marshall/marP.html [2.04.2017]
2. Wilson TD (2002) The nonsense of "knowledge management". Inf Res 8(1). Retrieved from http://informationr.net/ir/8-1/paper144.html [5.05.2005]
3. Nonaka I, Takeuchi H (1995) The knowledge-creating company: how Japanese companies create the dynamics of innovation. Oxford University Press, New York
4. Jashapara A (2014) Zarządzanie wiedzą (Knowledge management: an integrated approach). PWE, Warszawa
5. Davenport T, Prusak L (1998) Working knowledge—how organisations manage. What they know. Harvard Business School Press, Boston
6. Malhotra Y (2000) Knowledge management for [E-]business performance, information strategy. Exec J 16(4). Retrieved from http://www.kmbook.com/kmebiz/kmebiz.html [17.08.2004]
7. Fazlagić J (2006) Zarządzanie wiedzą. Szansa na sukces w biznesie. Gnieźnieńska Wyższa Szkoła Humanistyczno-Managerska, Gniezno

8. Tiwana A (2001) The essential guide to knowledge management; e-business and CRM applications. Prentice-Hall, Upper Saddle River
9. Toffler A (1990) Power shift. Knowledge, wealth, and violence at the edge of the 21st century. Bantam Books
10. Stefanowicz B (2012) Wiedza w interpretacji infologicznej. Zeszyty Naukowe Wydziału Informatycznych Technik Zarządzania WSISiZ "Współczesne Problemy Zarządzania" 1/2012
11. Fingar P (2007) The greatest innovation since BPM. Retrieved from http://www.bptrends.com/publicationfiles/SIX-03-07-COL-TheGreatestInnovationSinceBPM-Fingar-Final.pdf [2.12.2017]
12. Hansen M, Nohria N, Tierney T (1999 Mar–Apr) What's your strategy for managing knowledge? Harv Bus Rev 2(77):106–116
13. Leonard D (1995) Wellsprings of knowledge: building and sustaining the sources of innovation. Harvard Business School Press
14. Hislop D (2005) Knowledge Management in organizations. Oxford University Press, Oxford
15. Talisayon S, Talisayon A (2008) Tacit knowledge versus explicit knowledge. Retrieved from https://apintalisayon.wordpress.com/2008/11/ [16.07.2017]
16. Polanyi M (1967) The tacit dimension. Anchor Books, Garden City (NY)
17. Bennet A, Bennet D, Avedisian J (2015) The course of knowledge. A 21st century theory. MQI Press, Frost, West Virginia
18. Drucker P (1999) Knowledge-worker productivity: the biggest challenge. Calif Manag Rev 41(no. 2)
19. D'Aveni R (1994) Hypercompetition managing the dynamics of strategic maneuvering. The Free Press, New York
20. Ying CC (1967) Learning by doing—an adaptive approach to multiperiod decisions. Oper Res 15(5)
21. Gazzinga MS, Russell T, Senior C (2009) Methods in mind (Cognitive neuroscience). MIT Press, Cambridge
22. Szelągowski M (2013) Geneza dynamicznego zarządzania procesami biznesowymi. Kwart Nauk Uczel Vistula 4(38):41–56
23. Soosay C, Sloan T, Chapman R (2005) Developing organisational capabilities: a strategic approach to continuous innovation. In: 6th International CINet Conference, Brighton, England
24. Murray P, Myers A (1997) The facts about knowledge. Special report
25. Zerega, B. (1998). Art of knowledge management. InfoWorld [27.07.1998]
26. Wiig KM (1993) Knowledge management foundations: thinking about thinking—how people and organizations cerate, represent, and use knowledge. Schema Press, Arlington
27. Hammer M (1999) Reinżynieria i jej następstwa – jak organizacje skoncentrowane na procesach zmieniają naszą pracę i nasze życie (Beyond reengineering. How the process-centered organization is changing our work and our Lives). Wydawnictwo Naukowe PWN SA, Warszawa
28. Płoszajski P (2005) Zarządzanie (nie)wiedzą organizacyjną w Nowej Gospodarce. Podyplomowe Studium Zarządzania Wiedzą SGH
29. Probst G, Raub S, Romhardt K (2002) Zarządzanie wiedzą w organizacji. Oficyna Ekonomiczna, Kraków
30. Sopińska A, Wachowiak P (2006) Modele zarządzania wiedzą w przedsiębiorstwie. E-mentor 1(14)
31. Fahey L, Prusak L (1998) The eleven deadliest sins of knowledge management. Calif Manag Rev 40(3):265–276
32. Davenport T, Short J (1990) The new industrial engineering: information technology and business process redesign. Sloan Manag Rev 31(4):11–27
33. Davenport T (1995) The fad that forgot people. Fast Co Mag 1. Retrieved from https://www.fastcompany.com/26310/fad-forgot-people [12.03.2017]
34. Burlton R (2001) Business process management by Roger Burlton. Retrieved from https://www.slideshare.net/FreekHermkens/business-process-management-by-roger-burlton [14.08.2014]
35. Russell Records L (2005) The fusion of process and knowledge management. Retrieved from http://www.bptrends.com/publicationfiles/09-05%20WP%20Fusion%20Process%20KM%20-%20Records.pdf [8.04.2016]

36. Maier R, Remus U (2003) Implementing process-oriented knowledge management strategies. J Knowl Manag 4(7):62–74
37. Kim S, Hwang H, Suh E (2003) A process-based approach to knowledge-flow analysis: a case study of a manufacturing firm. Knowl Process Manag 10(4):260–276
38. Remus U, Schub S (2003) A blueprint for the implementation of process-oriented knowledge management. Knowl Process Manag 10(4):237–253
39. Maier R, Remus U (2002) Defining process-oriented knowledge management strategies. Knowl Proc Manag 9(2):103–118. https://doi.org/10.1002/kpm.136
40. Swenson K (2012) Case management: contrasting production vs. adaptive. In: Fischer L (ed) How knowledge workers get things done. Real-world adaptive case management. Future Strategies Inc., Lighthouse Point, Florida, USA
41. IBM (2014). IBM Business Process Manager V8.5.5 adds case handling and enhanced mobile UIs and IBM Business Monitor V8.5.5 provides more powerful analytics. Retrieved from https://www-01.ibm.com/common/ssi/cgi-bin/ssialias?infotype=an&subtype=ca&appname=gpateam&supplier=897&letternum=ENUS214-141 [20.07.2017]
42. Ultimus (2004) Adaptive discovery. Accelerating the deployment and adaptation of automated business processes
43. Ultimus (2004) BPM—simplified a step-by-step guide to business process management with The Ultimus BPM Suite
44. Kemsley S (2009). Dynamic BPM platforms. Retrieved from http://column2.com/2009/03/webinar-dynamic-bpm-platforms/ [3.04.2016]
45. Di Ciccio C, Marrella A, Russo A (2012) Knowledge-intensive processes: an overview of contemporary approaches? In: 1st international workshop on knowledge-intensive business processes (KiBP 2012) June 15th, Rome, Italy. Retrieved from http://ceur-ws.org/Vol-861/KiBP2012_paper_2.pdf [2.04.2016]
46. Marjanovic O, Freeze R (2012) Knowledge-intensive business process: deriving a sustainable competitive advantage through business process management and knowledge management integration. Knowl Process Manag 19(4):180–188
47. IEEE Task Force on Process Mining (2012) Process mining manifesto. Retrieved from http://www.win.tue.nl/ieeetfpm/doku.php?id=shared:process_mining_manifesto [02.04.2016]
48. Szelągowski M (2004). Szczegółowość identyfikacji procesów i działań w zarządzaniu dynamicznymi procesami biznesowymi. Zeszyty Naukowe "Studia i Prace" KZiF SGH w Warszawie, no. 49, 114–128
49. Gartner (2017) Magic Quadrant for Customer Communications Management Software. Technical Report ID G00298788; Published 26 Jan 2017
50. ISIS Papyrus (2017). Papyrus omni channel platform. Retrieved from http://www.isis-papyrus.com/e15/pages/ti/ti-omni-channel-E.html [10.07.2017]
51. Pucher M, Ruhsam C, Kim T et al (2014) Towards a pattern recognition approach for transferring knowledge in ACM. In: 2014 IEEE 18th international enterprise distributed object computing conference workshops and demonstrations
52. Sinha V (2017) BPM and big data—why it makes sense. Retrieved from http://www.bpminstitute.org/resources/articles/bpm-and-big-data-why-it-makes-sense [16.07.2017]
53. Szelągowski M (2015) Nowe metody zarządzania procesowego w ochronie zdrowia. E-mentor 5(62):40–48
54. Armistead C, Pritchard J, Machin S (1999) Strategic business process management for organisational effectiveness. Long Range Plan 1(32):96–106
55. Płoszajski P et al (2001) Zarządzanie wiedzą w Polsce. Bilans doświadczeń. Katedra Teorii Zarządzania Szkoła Główna Handlowa, Warszawa, 28
56. Jennings J, Haughton L (2002) Szybkość jako atut w biznesie: to nie duzi zjadają małych, ale szybcy opieszałych (Its not the big that eat the small … Its the fast that eat the slow. How to use speed as a competitive tool in business). MT Biznes, Warszawa
57. Abecker A, Papavassiliou G, Ntioudis S, Mentzas G, Muller S (2003) Methods and tools for business-process oriented knowledge management: experiences from three case studies. In: 9th international conference of concurrent enterprising, Espoo, Finland. Retrieved from http://imu.ntua.gr/sites/default/files/biblio/Papers/methods-and-tools-for-business-process-oriented-knowledge-management-experiences-from-three-case-studies.pdf [21.07.2017]

58. Remus U (2002) Process oriented knowledge management, concepts and modeling. Ph.D. thesis, University of Regensburg, Regensburg, Germany
59. Heisig P (2003) Wissensmanagement in industriellen Geschäftsprozessen (in German). Ind Manag 3. Gito-Verlag, Berlin
60. Gronau N, Müller C, Korf R (2005) KMDL—capturing, analysing and improving knowledge-intensive business processes. J Univ Comput Sci 4(11)
61. Jung J, Choi I, Song M (2007) An integration architecture for knowledge management systems and business process management systems. Comput Ind 58(2007):21–34
62. Di Ciccio C, Marrella A, Russo A (2015) Knowledge-intensive processes characteristics, requirements and analysis of contemporary approaches. J Data Semant 4(1):29–57. Retrieved from https://www.researchgate.net/profile/Claudio_Di_Ciccio/publication/269629902_Knowledge-Intensive_Processes_Characteristics_Requirements_and_Analysis_of_Contemporary_Approaches/links/576a501a08ae1a43d23bca3c.pdf [18.07.2017]
63. Buckley P, Carter M (2000) Knowledge management in global technology markets. Aplying theory to practice. Long Range Plann 33(2000):56−57
64. Vines R, Hall WP (2011) Exploring the foundations of organizational knowledge. Kororoit Institute Working Papers No. 3, 23–25
65. Dalmaris P, Tsui E, Hall WP, Smith B (2007) A framework for the improvement of knowledge-intensive business processes. Bus Process Manag J 13(2):279–305
66. Kemsley S (2010) Runtime collaboration and dynamic modeling in BPM: allowing the business to shape its own processes on the fly. Cut IT J 23(2):35–39
67. van der Aalst W, Dustdar S (2012) Process mining put into context. Internet Comput 16(1):82–83. Retrieved from http://wwwis.win.tue.nl/~wvdaalst/publications/p662.pdf [3.04.2016]
68. Chan N, Yongsiriwit K, Gaaloul W, Mendling J (2014) Mining event logs to assist the development of executable process variants. In: M. Jarke et al (eds), CAiSE 2014, LNCS 8484. Springer International Publishing Switzerland, pp 548–563
69. Handy Soft (2012) Dynamic BPM—the value of embedding process into dynamic work activities: a comparison between BPM and e-mail. Retrieved from http://www.bizflow.com/system/files/downloads/HandySoft%20-%20Dynamic%20BPM%20White%20Paper_0.pdf [2.04.2016]
70. Senge P (1990) The fifth discipline. The art and practice of the learning organization. Currency Doubleday, New York
71. Mikuła B (2001) W kierunku organizacji inteligentnych. Antykwa, Warszawa
72. Lassey P (1998) Developing a learning organization. Kogan Page, London
73. Argyris C, Schon D (1978) Organizational learning: a theory of action perspective. Addison-Wesley, Reading, MA
74. Gladstone B (2004) Zarządzanie wiedzą (From know-how to knowledge). Wydawnictwo PETIT, Warszawa
75. Tsang E (2016) Organizational learning and the learning organization: a dichotomy between descriptive and prescriptive research. Hum Relat 50(1):73–89
76. Pfeiffer J, Sutton R (1999) The knowing-doing gap: how smart companies turn knowledge into action. Harvard Business School Press, Boston
77. Perechuda K (2004) Drgająca dysfuzja wiedzy w przedsiębiorstwach sieciowych. Retrieved from http://efektywnosc04.ae.wroc.pl/Referat/art03.pdf
78. Płoszajski P (2004) Organizacja przyszłości: przerażony kameleon. W kierunku nowej filozofii zarządzania. Retrieved from http://www.allternet.most.org.pl/SOD/Heterarchia%20prof._Ploszajski_-_Organizacja_przyszlosci.pdf [22.04.2017]
79. Taylor C (2012) Reunifying knowledge and business process management. Retrieved from http://citeseerx.ist.psu.edu/viewdoc/download?doi=10.1.1.225.9570&rep=rep1&type=pdf [18.07.2017]

Chapter 4
The Implementation of Dynamic Business Process Management

Abstract The implementation of process management requires, first and foremost, the introduction of changes to the fundamental principles of management within the organization. It is insufficient for the owner or the CEO to decide on the development of new infrastructure, the purchase of new computers and new software, or even the introduction of comprehensive personnel training. Process management requires the use of proven methodologies and the skillful implementation of their supporting IT systems with a view to changing the organizational culture itself. The aim of this chapter is to present the results of studies on the higher efficiency of implementing and executing dynamic business process management, as compared with its traditional counterpart. The chapter then goes on to describe the changes in defining goals and preparing descriptions of business processes throughout the entire process lifecycle, which are the result of the requirements of business within the knowledge economy. The chapter provides an overview of the research and its conclusions, which point to the necessity of introducing changes to the principles of describing and presenting business processes. To conclude, the chapter presents the principles of implementing dynamic business process management, including the use of process exploration techniques, as well as the integration of process management with knowledge management. Drawing from the author's personal experience, which stems from cooperation with diverse research teams executing commercial implementation projects in the years 2006–2017, as well as based on a review of relevant scholarship in the field, the chapter points to those fields, in which the use of a dynamic approach does not offer virtually any benefits, as well as those fields, in which it provides significantly better results than traditional process management or is a necessary condition for process management to provide any results at all.

Keywords Business process management (BPM) · Dynamic business process management (dynamic BPM) · Business process model and notation (BPMN) · Case management (CM) · Process mining · Process automation · Knowledge management (KM) · Process-oriented knowledge management (pKM) · Knowledge-intensive business processes (kiBPs)

© Springer Nature Switzerland AG 2019   137
M. Szelągowski, *Dynamic Business Process Management in the Knowledge Economy*, Lecture Notes in Networks and Systems 71,
https://doi.org/10.1007/978-3-030-17141-4_4

4.1   Introduction

The implementation of process management requires, first and foremost, the introduction of changes to the fundamental principles of management within the organization. Process management requires the use of proven methodologies and the skillful implementation of their supporting IT systems, with a view to changing the organizational culture itself. The aim of this chapter is to present the results of studies on the higher efficiency of implementing and executing dynamic business process management, as compared with its traditional counterpart. By answering the following research questions: Is it possible to describe dynamic business processes, and *Is it possible to perform simulation and optimization studies on dynamic processes, including unpredictable processes?*, the chapter describes the practical possibilities offered by dynamic business process management combined with new ICT technologies and new analytical and simulation techniques. Implementing traditional process management without implementing IT systems supporting the performance of business processes is very difficult, if not downright impossible, in practice. Such an attempt would be burdened with all of the drawbacks associated with implementing quality control systems on the basis of documentation prepared and updated in paper form. It would require the constant, extremely burdensome, slow, and risk-prone copying and distribution of new or updated documentation (upon prior authorization), as well as the withdrawal of obsolete documentation (again, upon prior authorization). Instead of raising management efficiency, it could result in a drop in efficiency due to the imposition of additional, inhibiting bureaucracy.

The implementation of dynamic business process management without IT support is impossible. The implementation of the 2nd principle of dynamic BPM, which has been defined in Chap. 2, without the possibility to monitor actions undertaken in the organization on an ongoing basis, including deviations from the standard process "as of today," could result in loss of control and, in effect, in rising organizational chaos. The real-time collection and analysis of large datasets consisting of hundreds of thousands or even millions of records, which additionally change over time, is downright impossible without the use of IT systems [1, pp.11–12, 89–91] . For this reason, the implementation of process management is strictly tied to implementing its supporting IT systems. However, implementations of process management have to meet additional goals and requirements in the knowledge economy in comparison with those set in the industrial economy in regard to implementations of MRP II, ERP, or CRM systems. This is the result of, first and foremost, different goals of implementing process management. In the case of implementations of Material Requirements Planning (MRP), Manufacturing Resource Planning (MRP II), and integrated Enterprise Resource Planning (ERP) systems, as well as Business Intelligence (BI) components, which are increasingly more often essential thereto, the aim was to use IT tools with a view to raising efficiency (in particular—production efficiency) or the improvement of control functions within the organization. Such attempts usually had the form of one-off (though sometimes multiyear) projects with the aim of outfitting the organization with particular functionalities or IT services.

The aim was not to optimize business processes or the operations of the organization as such. On the contrary: oftentimes, such implementations resulted in the organization and the processes becoming subordinated to the requirements of the specific IT tool itself. However, despite no rise in the efficiency of business processes, or even a drop thereof (e.g. by the bureaucratization of operations, which resulted in a bloated organizational structure and an overcomplication of processes), the fact of using an integrated IT system alone has resulted in large enough savings and benefits that at the end of the day, the implementation increased the efficiency and transparency of operations. The rise in efficiency resulting from the implementation of an IT system did not require changes to management, more engagement on the part of the employees, or the optimization of processes, etc.

Implementations of process management set forth different goals for the organization. Not only do they require the skillful use of supporting IT systems, they primarily require changes to the fundamental principles of management within the organization. It is insufficient for the owner or the CEO to decide on the development of new infrastructure, the purchase of new computers and new software, or even the introduction of comprehensive personnel training. Process management requires the use of proven methodologies and the skillful implementation of their supporting IT systems, with a view to changing the organizational culture itself. All of the employees (or the vast majority thereof) must be aware of and adhere to the principles of operation pertaining to business processes on a daily basis, as well as be aware of and adhere to the principles of adapting processes to the changing conditions of business. For this reason, it is usually not technical problems, but the hidden convictions or the inconsequential actions of the management, as well as the silo mentality of the organizational culture, which turn out to be the largest threat that implementations are faced with. At the same time, process management does not encompass the entire management within the organization. As the structure of the Balanced Scorecard demonstrates, business processes are just one of the perspectives of approaching management within the organization. In effect, process management must be harmonized with other management instruments.

4.2 The Implementation of Traditional Process Management

The implementation of process management requires organizations to introduce principles of management, which will center all of the operations of the organization on the goals of the fundamental process, or the process which results in the creation of the organization's main products and services [2, pp. 119–128; 3]. In order to fulfill this goal, it is essential to both eradicate the silo mentality of the organizational culture, as well as to rationalize processes with a view to subordinating them to the fundamental process. As has been discussed in Chap. 1, the last 10–15 years saw

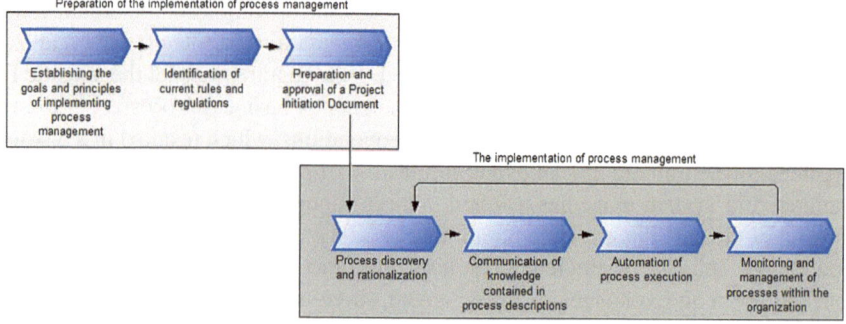

Fig. 4.1 Standard steps of implementing traditional process management. *Source* Author's own elaboration

the appearance of a dozen or so different concepts and methodologies,[1] as well as several hundred BPMS, CMS, and ERP IT systems supporting the implementation and adoption of process management.[2] The standard steps of implementing traditional business process management, which are used in most cases, are presented on Fig. 4.1.

Traditional implementation of process management is usually divided into two stages:

I. Preparation of the implementation of process management—this stage is executed once;
II. The implementation of process management—this stage is executed multiple times.

With each subsequent execution, the second stage encompasses the group of business processes agreed upon with the management of the organization and heads of the implementation team. In accordance with agile methodologies (e.g. Agile Project Management [4] or Scrum [5], the outset of this stage should include an analysis of the achieved results and an overview of the list of subsequent process groups intended for implementation, update, or improvement. This should be followed by an update of the implementation plan and the entire stage should be executed again for the selected process group. In accordance with the Deming cycle, this process never ends [6, pp. 42–44]. This is not only the result of expanding the implementation to encompass additional processes within the organization, but primarily the result of ongoing changes to business requirements, as well as the constant development of employees and the rising process maturity level of the organization itself.

---

[1]For example, BPM Framework (Process Renewal Group), Customer Expectation Management Method (CEMM) (BM Group), Rapid Application Development and Systems Integration (BIZFLOW), IBM Playback Methodology (IBM).

[2]The single Software Advice website compares as many as 134 ERP systems (http://www.softwareadvice.com/erp/), as well as 17 BPMS systems (http://www.softwareadvice.com/bpm/).

Tables 4.1 and 4.2 present the main actions undertaken in the course of executing stages I and II of implementing process management.

One should stress the fundamental significance of the Establishing the goals and principles of implementing process management stage. Lack of agreement in this regard, or the imposition of such goals by the management of the organization itself, point to a lack of understanding in regard to the principles of process management, e.g. the necessity to empower the employees themselves. This, in turn, might point to the lack of sufficient process maturity within the organization, preventing it from successfully implementing process management, which would result in the implementation having to be halted.

However, even in such circumstance the implementation team could still continue with process discovery or even partial process automation.

4.2.1 Process Discovery and Rationalization

The main goal of the Process discovery and rationalization stage is to define the ongoing performance of processes within the organization and to develop improvements which would be beneficial from the perspective of the organization as a whole. In accordance with the standards of traditional process management, the identification of processes should be executed with the use of a top-down approach, that is, identification should begin with the most complex processes (megaprocesses) and turn to more detailed processes later on. This sequence is the result of the necessity of focusing all operations, including rationalization and optimization, on the efficiency of the fundamental process, or the process thanks to which or for which the organization exists to begin with [7, 8].[3] In effect, it is essential—on the level of process maps to start with—to establish which process is the fundamental process, which support processes are the most crucial thereto, what this support entails, and how it can be intensified. Since in accordance with the Process Relevance Criterion different groups of processes may be modeled and described on different levels of detail, it is possible to prepare detailed diagrams for selected processes alone, e.g. processes which are the most influential in terms of the results or the competitive advantage of the organization, as well as processes with the highest risk factor [9]. Usually, the Process discovery and rationalization stage results in the creation of:

• process models for current processes within the organization (as is);
• process models for improved, target processes (to be);
• analyses and risk mitigation plans pertaining to disrupting the interests of particular individuals and groups of employees as the result of improving processes.

In practice, during process modeling sessions a model is created to account for rationalizations suggested by employees in regard to current (as is) processes. Pro-

---

[3]In the case if business its simply the process, thanks to which the organization makes profit (earns money). In the case of non-profit organizations, the fundamental process is the one for which the organization has been created to begin with.

Table 4.1 The main actions undertaken in the course of executing stage I—preparation of the implementation of process management

Stage	Goal	Participants	Actions	Products
Establishing the goals and principles of implementing process management	Minimization of the risk of misunderstanding the basic principles of process management	The organization's top management	Defining and agreeing upon the goals of the implementation Defining the principles of implementation, including the level of engagement of the organization's top management	Official "implementation goals" document
Identification of current rules and regulations	The effective use of knowledge and tailoring the implementation to the process maturity of the organization	The organization's top management, senior management	Overview of current internal (procedures, statutes) and external (e.g. legal requirements, licenses) rules and regulations Defining the process maturity level of the organization Overview of IT resources, with a focus on maintaining ongoing support for business processes	Initial version of the organization's process map Description of the architecture of the ICT systems within the organization
Preparation of the project	Preparing a Project Initiation Document (PID), including a workable implementation plan accepted by the participants of the implementation	The organization's top management, senior management, process owners and process managers	Preparation of a process description standard. Preparation and approval of a process map Selection and preparation of training sessions for process owners and process managers and the implementation team Preparation and approval of a formal Project Initiation Document (PID)	Process map for the organization Project Initiation Document (PID)

Source Author's own elaboration

Table 4.2 The main actions undertaken in the course of executing stage II—the implementation of process management

Stage	Goal	Scope	Actions	Benefits
Process discovery and rationalization	Preparation and approval of process descriptions for the selected group of processes. The optimization of the described processes and the preparation of a standard process, understood as the most efficient performance of a given process, which will serve as the effective pattern of execution for process performers	Primarily the fundamental process	Goals and goal indicators for processes should be defined prior to initiating process modeling. "Aimless" processes should not be modeled. Preparation and approval of process models for selected processes in their current (as-is) and improved (to-be) versions in accordance with the organization's overall process model. In the modeling stage, processes are usually rationalized, understood as the elimination of errors which were not identified prior to process modeling even by employees with years' worth of experience, and which are now obvious to all in the modeling stage. Process diagrams should be unambiguous and understandable not just to process modelers, but primarily to all potential users, who will be obligated to make use of the information contained in process models	Higher efficiency (e.g. reducing the time of execution, reducing the costs and resource consumption) and lower risks thanks to the optimization of the standard process. Gathering and disseminating knowledge within the organization (e.g. using process description during training sessions for new employees or in communication with clients or business partners)

(continued)

Table 4.2 (continued)

Stage	Goal	Scope	Actions	Benefits
Transfer of knowledge contained in process descriptions	Dissemination of knowledge included in process models and descriptions. Verification and creation of knowledge with the participation of managers and experts responsible for the processes	The Fundamental Process, but also support and management processes Primarily the employees engaged in the execution and controlling of process execution	The publication of process models and descriptions for all authorized people of interest, with a focus on the employees, but in justified cases also subcontractors, suppliers, or even the clients themselves The possibility of holding two-way communication with managers and experts responsible for particular processes with a view to: – verifying explicit knowledge available in the form of process models and descriptions (taking in questions and proposals for improvement), – using the tacit knowledge of the managers and experts	Ongoing access to the knowledge of the organization The possibility of using and objectively evaluating the knowledge and the engagement of employees and experts The elimination of losses resulting from the unfamiliarity with and the non-application of standard in ongoing operations Reducing the workload thanks to access to standard documents during process execution (e.g. a repository of document templates)

(continued)

Table 4.2 (continued)

Stage	Goal	Scope	Actions	Benefits
The automation of process execution	Raising efficiency through work automation	The Fundamental Process, but also support and management processes Primarily processes with a high frequency of performance or requiring strict repetition	Iterative preparation and implementation of process-driven applications, that is, applications in which the logic of operation is subservient to process diagrams Changes to process diagrams and the creation of new process scenarios result in a change in the application's operation.	Higher efficiency (e.g. by reducing the workload and limiting resource consumption) and lower operational risks thanks to the creation of conditions in which the organization is ensured to operate in accordance with the standard (the best possible) processes The pace of tailoring applications supporting the execution of processes to changes to the processes in the organization
Monitoring and management of processes within the organization	Managing business processes, not just their results	Primarily the Fundamental Process Support processes with the highest risk for the Fundamental Process Management processes which are strategic in the organization	Ongoing monitoring and management of automated processes (standard control mechanisms) Monitoring and correcting the execution of the implementation plan, e.g. by the ongoing analysis of priorities and the selection of subsequent groups of processes to undergo discovery, communication, and automation	Reduction in monitoring and controlling costs thanks to the automate collection of process results and performance indicators Reduction or elimination of losses thanks to the automatic, advance signaling of threats an alarms The possibility of fast, competitive reaction to threats resulting from changes in the market thanks to ongoing access to reliable data on the performed processes

Source Author's own elaboration

cess rationalization is understood as the practice of fixing errors or introducing ("obvious") improvements, the necessity of which prior to process modeling was not realized even by employees with years' worth of experience, performing the processes and having no comments in regard to their performance—even unofficially. However, in the process modeling stage it often turns out that "it is impossible for us to work in this way." This realization leads to the necessity of immediately fixing the inoptimalities or errors, which have not been identified thus far, and which became obvious to all in the modeling stage. In this case, a model rationalized in accordance with "common sense" usually replaces the model of as-is processes.

Process diagrams prepared in the Process discovery and rationalization stage must be unambiguous and understandable not just to the modeling team, but to all potential users who will make use of the knowledge contained in the process models. For this reason, it is essential to choose the right form and notation for business process descriptions [10]. Because notations often have the form of broad collections of symbols and rules (e.g. the specification of the BPMN 2.0.2 notation amounts to 532 pages [11]), one should be selective in their choice of the symbols and rules of modeling used in model diagrams. The criterion for selection should rest in making sure that the models clear to their future users. Too small a selection of symbols may result in problems with underlining the business logic behind the model, too broad a selection—with being able to read and understand the prepared diagrams, and in result: with using the knowledge contained therein. In practice, when using e.g. BPMN notation it is sufficient to select about 20 symbols to model most of the business process within the organization in a manner which is clear and understandable to recipients without broad knowledge and experience. Of course, the larger their knowledge an experience, the more this selection can be broadened.

One good practice in the Process discovery and rationalization stage is to use the following as a source of knowledge about the actual course and the rules of monitoring processes in the organization:

- formal documentation (rules and regulations, procedures, diagrams, decisions, report patterns, decision tables, quality manuals, etc.);
- the informal and often tacit knowledge of line workers (working, "private" Excel spreadsheets or Gantt diagrams, unofficial rules of communication and control, or—in general—daily experiences);
- the actual document and information flow—identified on the basis of existing documents and confirmed in the course of e.g. non-participant observation [12, pp. 343–360].

4.2.2   The Transfer of Knowledge Contained in Process Descriptions

The participants of the Process discovery and rationalization stage usually consist of the implementation team and experts (internal and external) invited to model specific process groups. Knowledge collected in the course of process discovery is disseminated throughout the organization, with a view to its:

1. **Verification**—for this reason, knowledge should be disseminated by means of two-way communication, which not only enables employees to gain access to knowledge from process descriptions in the form of a knowledge repository, but also to ask questions, make comments, and propose improvements. The latter should be assessed by the experts who authored or co-authored the process descriptions in question. They, along with the process owners themselves, are responsible for keeping process descriptions up to date, including the adaptation of processes to changing technologies, conditions of concluding business, culture, or organizational knowledge;

2. **Use in ongoing operations**—for this reason, process descriptions should be clear to a broad range of employees. One good practice is for the knowledge to be accessible directly in the place of performance itself. Accessing a knowledge repository with process descriptions should not require the employees to interrupt work, abandon their workstation, or install or launch additional applications. This could discourage them from:

   a. keeping up to date with the current knowledge of the organization (which could be beneficial to the bottom line);
   b. sharing knowledge on more effective or more innovative process executions, which is available locally to select employees.

In the Transfer of knowledge contained in process descriptions stage, of help are different ICT technologies, knowledge depositories in the vein of Wikipedia, social applications, messaging apps, and process portals—also available in mobile form. Their use enables:

- avoiding losses resulting from lacking knowledge on standards and mistakes pertaining to ongoing operations;
- the broad reporting of ideas and improvements—the ongoing improvement of processes;
- sharing knowledge and experiences within the organization;
- the quick dissemination of acquired knowledge in the form of e.g. databases (lists) of best and worst practices;
- reducing the workload thanks to access to templates of standard documents drafted in accordance with process models during process execution;
- the use of processes descriptions as ongoing documentation in the course of training sessions for new employees or business partners.

4.2.3 The Automation of Process Execution

The main goal of automating process execution is raising efficiency as the result of reducing the workload and raising the pace of document management and workflow within the organization. One of the crucial benefits of automating process execution is the creation of competitive advantage as the result of raising the flexibility of management. This is primarily the result of reducing the necessary time spent by members of IT on adapting process-driven applications to the changing requirements of business, which allows for:

- reducing the time between updating a process model and implementing the change in a system supporting process execution;
- reducing the costs of updating IT systems.

The automation of process execution also allows for raising work efficiency, as it makes work more fluid regardless of the functional structure of the organization. This, in turn, allows for the elimination of simple losses resulting from e.g.:

- the misharmonization of actions and the improper information flow between different organizational silos (e.g. waiting for events that have already taken place or failing to realize the necessity of undertaking actions in the course of a process, the performance of which has already been transferred to another organizational unit);
- unnecessary repeated input of the same data (e.g. messaging app, e-mail, ERP system, scanning system, e-mail/fax or multiple forms connected with different databases);
- direct presence to perform tasks which could be completed remotely: with the support of IT systems or mobile applications (e.g. simple organizational and financial processes).

The automation of process execution should not be executed in the form of an instant, one-off replacement of the transactional (MRP II, ERP, or CRM) and ICT (e-mail, fax, messaging software, social media applications) systems within the organization. Such an approach would, in the short term, result in:

- the considerable cumulation of changes in the IT systems that employees must come to terms with, which usually results in their reluctant, or even adverse attitude toward the implementation in question;
- a high risk of the improper operation and interoperation of new or newly reconfigured systems, which might threaten the continuity of ICT support in the organization;
- high one-off costs of changing the systems, as well as usually additional high costs of changing the ICT infrastructure.

A much more efficient approach lies in systems integration, with a particular focus on organizing and integrating sources of data. In the Process monitoring and management stage, it allows for analyses encompassing a much broader spectrum of cohesive data than merely the analysis of chronologically organized actions. The integration of process-driven applications with existing transactional systems allows for the minimization of the time, costs, and risks involved with the project of implementing a new solution. The implemented process application usually replaces part of the functionality of an existing transactional system (e.g. enables simultaneous

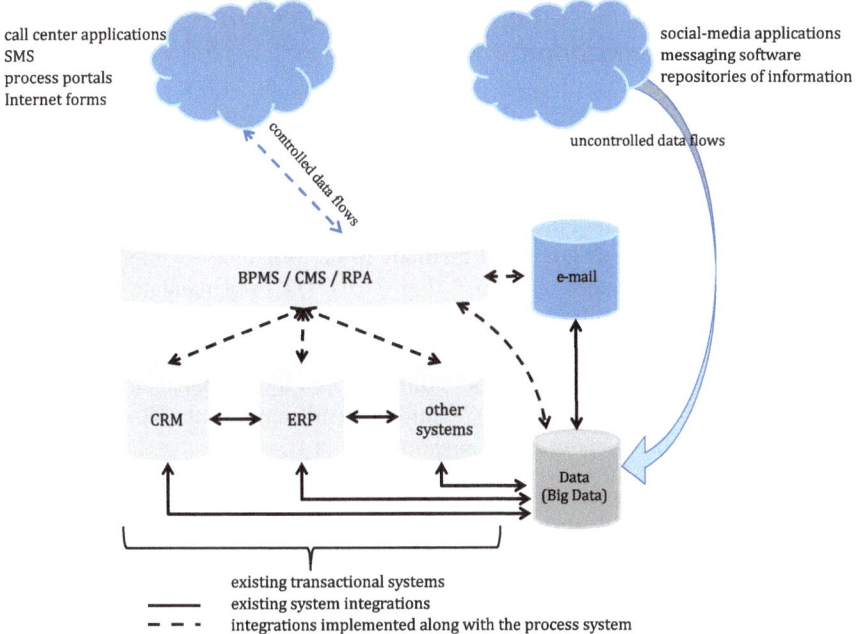

Fig. 4.2 Sample (demonstrative) architecture of IT systems during implementation of process management. *Source* Author's own elaboration

access to multiple variants of processes in new channels of communication with the organization) or provides entirely new possibilities of performing business processes, which were not available before (e.g. Internet sales, group purchases) (Fig. 4.2).

Another good practice is to begin the automation of process execution not from complex processes, which are fundamental for the organization's bottom line, but from simple document management or workflow processes, the successful automation of which is privately important and desirable to the employees. Examples of uncomplicated processes, whose goals further motivate the employees, are e.g. the process Acceptance of a leave of absence request, the process Acceptance of a request for ICT hardware (computer, phone, etc.), or the process Advance payment clearing. Such changes make the employees more engaged in the use of process-driven application due to the fact that they have a personal interest in the effective performance of the processes in question—which minimizes their usually reluctant attitude to new solutions. At the same time, for the implementation team and the ICT team, this is a good opportunity to:

- check the proper functioning of the entire ICT architecture (including the integration of administrative, transactional, and process-driven systems);
- verify the efficiency of creating and testing process-driven applications;
- verify the efficiency of the training sessions held in the course of implementation.

4.2.4   Traditional Business Process Management in the Organization

Initially, business process management in the organization has the goal of monitoring the course of the implementation of process management and quickly reacting to potential risks and threats thereto, in order to minimize them. Ultimately, however, its primary goal is to monitor the efficiency of process execution and to take steps with a view to optimizing their goals and methods of performance—within the 1st and 2nd learning loops described in Chap. 3 (Fig. 3.16). The Implementation of process management stage is usually executed as part of process management itself, which allows the initiation of the ongoing, practical improvement of business processes not during performance itself, but at specific times chosen by the management [13].

The critical success factors (CSFs) pertaining to the implementation of traditional business process management in organizations are [14]:

- the harmonic combination of strategies and processes within the organization;
- the conviction and ongoing commitment of the organization's management;
- breaking down walls between organizational silos, which results in focusing processes on clients;
- preparing a process architecture enabling the identification of processes and the efficient management of changes;
- the management of the ICT architecture, which allows for the successful dissemination of knowledge and the implementation of process-driven applications;
- the implementation of a control system to monitor and allow for prognostics of the performed processes; this project should encompass strategic performance indicators and indicators used and understood in specific positions in the workplace;
- the ongoing monitoring of performance and agile management of the implementation project, including the organization's level of process maturity.

As has been demonstrated in Chap. 1, traditional implementations of process management in the organization do not eliminate, but even strengthen the division between the decision on the method of performing work and the performance itself, which results in:

- the inability to individualize processes,
- a top-down approach to introducing changes,
- no possibility of holding limited experiments on a broad scale,
- the failure to follow changes in customer requirements and the competitive environment,
- blurring the responsibility for the results of process execution,
- using just a small part of the entire knowledge of the organization,
- insufficient support for knowledge management or even direct conflict with the concept of knowledge management.

Implementing traditional process management does not empower the organization to achieve and maintain its competitive advantage in the knowledge economy.

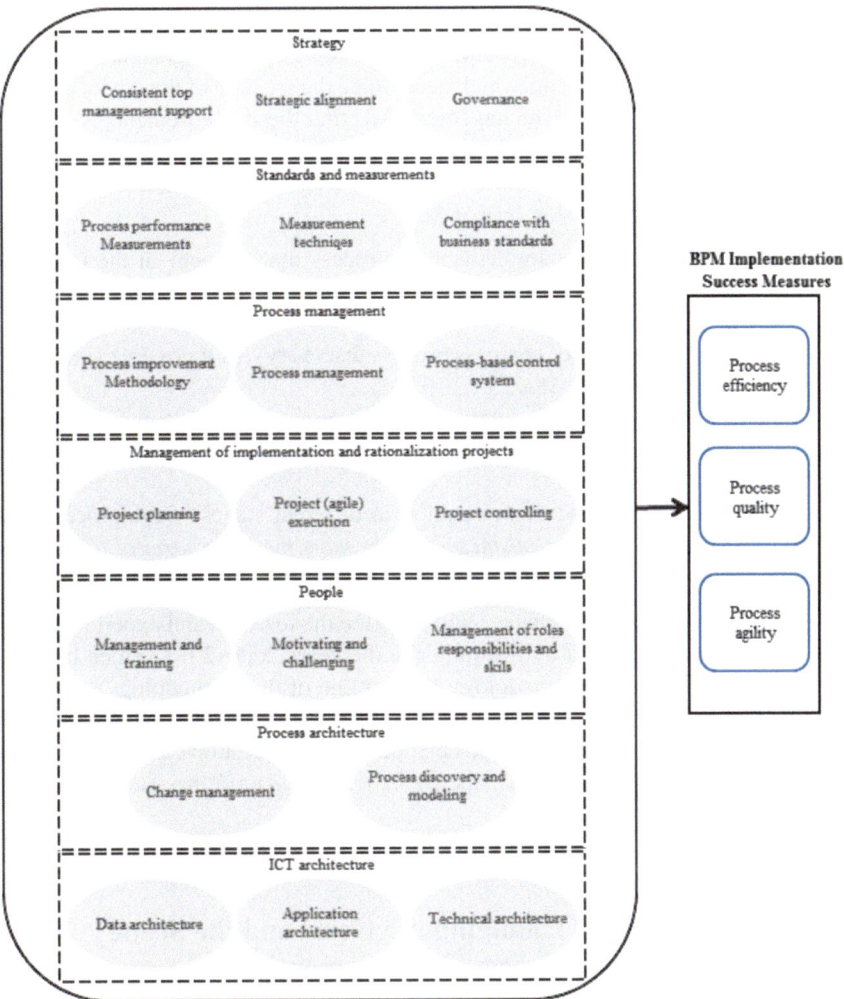

Fig. 4.3 Critical success factors (CSFs) behind implementing traditional business process management in organizations. *Source* Author's own elaboration, on the basis of Dabaghkashani et al. [14, p. 727] (This is different from Michael Rosemann and Jan vom Brocke: *"Each of the six core elements represents a critical success factor for Business Process Management. …. Our model distinguishes six core elements critical to BPM. These are strategic alignment, governance, methods, information technology, people, and culture"* Rosemann and vom Brocke [15])

Process management encompassing a mere 20–30% of the business processes of the organization along with slow and limited knowledge management are clearly unable to meet the requirements which necessitate the creation of the 4th, or even the previous 3rd wave of process management. Meeting these requirements is possible thanks to the implementation of dynamic business process management. However, this proves impossible without introducing certain changes to the methodologies of implementation, encompassing (Fig. 4.3):

1. Changes to the goals of implementing process management in the organization—*What sort of goals should be set before (dynamic) process management in the knowledge economy?*
2. Changes to the method of business process discovery—*Is it possible to describe dynamic business processes?*
3. Changes to the method of holding simulation research in the context of business processes—*Is it possible to research dynamic processes, including unpredictable processes?*
4. Changing forms of process description in different stages of the process lifecycle—*Should a dynamic business process have the same form of description throughout its entire lifecycle?*

The following part of the chapter will describe the research and experiences pertaining to projects executed by the author in the years 2006–2017, which have led to the formulation of the methodology (the outline of the methodology) of implementing dynamic business process management in organizations. This part of the chapter will provide an answer to the detailed research questions, as well as—once again—the question pondered in Chap. 2: *Is it possible to unify process management with case management?*

4.3   The Necessity of Changing the Goals and the Scope of Implementing Process Management

As has been discussed in the above section, in the traditional, static approach to implementing process management, the goal of implementation is usually to "optimize" or to "raise the efficiency" of process execution. Implementing dynamic BPM requires us in practice to take into account the fact that in the knowledge economy, intellectual capital is more important than tangible assets, and in consequence—to include in the goals and methods of implementing process management its integration with knowledge management as the foundation of the organization's long-term competitive advantage. In other words, the goal can be formulated as "raising efficiency and using knowledge on an ongoing basis," "optimizing the operations and management of the entire available knowledge," or similarly, depending on the strategic priorities of the organization [16]. What should result from the formulation, in a way which is obvious to all in the organization, is that the goal is not the one-off uncovering of knowledge in the Process discovery and rationalization stage, which will then be updated during subsequent reviews or certification audits. Rather, starting in the Process discovery stage, the team implementing dynamic process management should, with the consistent support of the organization's management, communicate to the entire personnel that within process management, one should on an ongoing basis:

- make use of the available knowledge;
- create and reveal new knowledge;
- share knowledge, so that the organization will gain benefits faster than the competition.

In effect, on the level of rules or strategies of management we are dealing with three dimensions of expanding the goals of implementing traditional process management, such as:

1. (broader) incorporation of the knowledge and engagement of all knowledge workers instead of just the views accepted by the management;
2. (more complex) incorporation and integration of knowledge management with process management;
3. (quicker) systemic introduction of ongoing knowledge management and process management instead of one-off or periodic updates.

One should underline that we are speaking of revealing knowledge that has already been used in practice in the course of performing business processes. In other words, the knowledge in question can be objectively measured, and the results if its use can be presented with the use of indicators pertaining to exploratory analyses of the performed processes. In this way, thanks to process exploration or analyses of big data, dynamic business process management has gained a tool which enables the quick, objective analysis of the performance of expanded goals. We can also point to specific process performers who are responsible for the creation and use of new knowledge. In effect, knowledge management is no longer limited to managing knowledge created by the management, but encompasses the knowledge of a broad range of process performers, whom the organization is able to evaluate and reward as individuals.

When answering the fundamental question: *What sort of goals should be set before (dynamic) process management in the knowledge economy?*, it should be said that the goal of implementing dynamic BPM is:

- raising the efficiency of process execution (as in traditional business process management);
- the ongoing optimization of the use of the entire knowledge of the organization and its management—as the result of the creative and innovative use of knowledge in the course of process execution.

By expanding the goal of implementing process management, organizations gain two additional benefits:

- a much broader use of the organization's intellectual capital with the aim of creating value;

- the possibility of quicker, objective evaluation of the results of using knowledge with the aim of maximizing the pace and the efficiency of knowledge management.

Expanding the scope of business process management (understood as dynamic BPM) in comparison with traditional business process management arises from the much broader scope of business processes that can be managed. It is no longer—as in the case of traditional process management—just 20–30% of the processes of the organization, but all of the processes therein, even unstructured, unpredictable ones. As has been demonstrated in Chap. 1, as recently as five years ago the modeling of dynamic processes has been considered impossible [17] (Fig. 4.4).

At the same time—in accordance with the expanded goals of implementing process management presented above—within the knowledge economy most of the processes in the organization are knowledge intensive, or processes whose performance requires up-to-date knowledge and the course of which both verifies current knowledge and creates new knowledge. This requires us to supplement the principles of implementing process management to account for dynamic processes, but also (as a special case)—to account for static processes as well. In effect, this requires the consistent change in the method of:

- identifying and describing business processes;
- researching and analyzing completed and ongoing business processes;

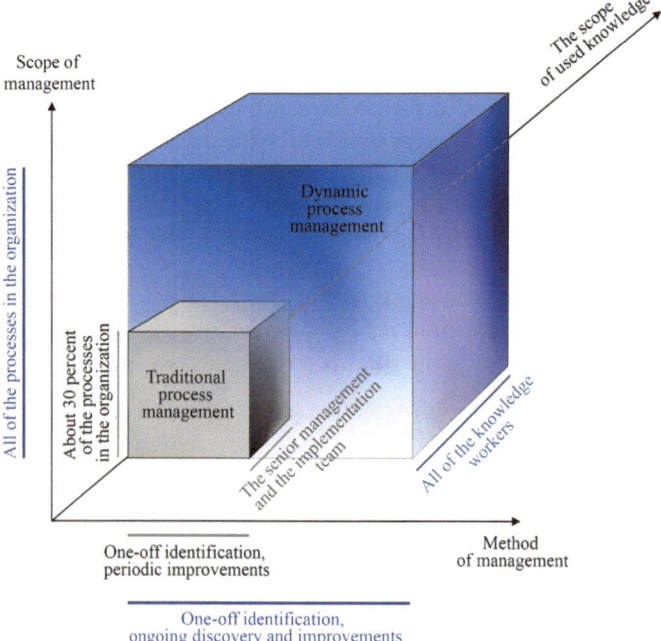

Fig. 4.4 Expansion of the scope of implementations within dynamic process management. *Source* Author's own elaboration

- automating the performance of business processes;
- monitoring the performance and analyses of business processes.

4.4 Changing the Method of Business Process Description

There have been multiple attempts at standardizing the principles of process description through introducing notation describing business processes. The best-known notations are as follows:

- BPMN (Business Process Model and Notation, OMG);
- BPMS (Business Process, BOC);
- EPC (Event-driven Process Chain, Software AG);
- UML (Unified Modeling Language, OMG);

In 2014, OMG also introduced Case Management Model and Notation (CMMN) devoted to case management [18]. At present, the most popular notation by far is the 2.0.2 version of Business Process Model and Notation (BPMN) [11]. It allows for the modeling of static processes, as well as the modeling of dynamic processes to a large degree.

4.4.1 Traditional, Static Process Modeling

In traditional process management, process descriptions usually contain process models in the form of diagrams defining both the course (the logic) of the process, as well as the responsibilities of its particular participants, the documents used, the essential data, IT systems supporting process execution, its indicators, etc. They contain all of the information on performing the work known to the organization. They are prepared by a team which discovers and designs processes in a way which is detached from process execution itself, that is, without being aware of the client and the specific context of performance, and usually without reflection on the experience of the individuals who are designated to perform the process. For this reason, a process diagram or a process description include all of the possible and probable actions and objects which could appear in the course of the process. Among the three elements comprising our knowledge on a given process, process descriptions usually include information (I) and general data on the context (C) of process execution. Figure 4.5 depicts a diagram of an example process Decision consultation. It contains all of the roles which should (or perhaps have to) participate in making the decision.

This process certainly did not foresee all of the requirements and possibilities, so the improvement process will require its constant supplementation. The management will be made aware of these necessary improvements in the course of analyzing the results of suboptimal or even wrong decisions. The expanded process will become

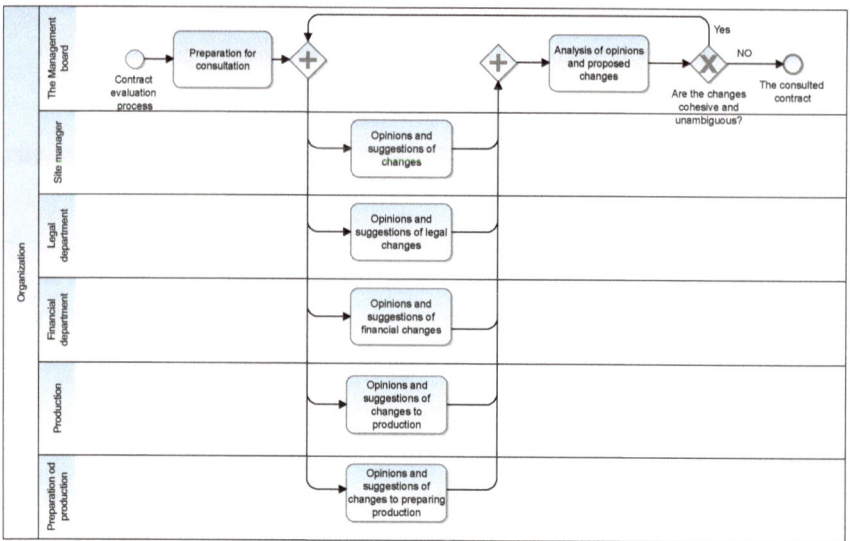

Fig. 4.5 Example of static business process modeling. *Source* Author's own elaboration

increasingly less transparent and will start to include "just in case" provisions, usually unnecessarily so, as well as call upon disinterested individuals for no good
reason to oversee decisions in specific situations. Because the process is designed
in a way which is detached from specific performances, in order to minimize risk
it must include all possible scenarios. The model may be improved by replacing
AND decision gates with OR decision gates. This enables the consultant to select
the participants of the consultation. However, it does not allow for the choice of the
sequence of the consultations itself (this is an entirely different process with even
greater complexity), nor for the dynamic, context-dependent expansion of the group
of consultants. This would require the complete overhaul of the entire process. Perhaps it would be wiser to model decision-making processes in response to specific
events? Figure 4.6 provides an example of static business process modeling in the
BPMN notation on the basis of the context of events.

The first gate—regardless of whether it is a XOR gate, an OR gate, or an event-
based gate—has the function of distinguishing between the types/contents of making
a decision (e.g. signing a contract). A single process path (or plural process paths) is
entered—one which corresponds to the type of required consultation before making
the final decision on signing the contract. Additionally, the consultation process
may be terminated due to time constraints. This would require making the final
decision on signing the contract regardless of the unfinished state of consultations
(due to the timed event Consultation deadline). Regardless of which decision gate
we use, we end up with a process which requires us to predict and model all of
the decision scenarios prior to its execution. Even when all of the scenarios are
modeled in an optimal manner, the number of scenarios will increase with time,

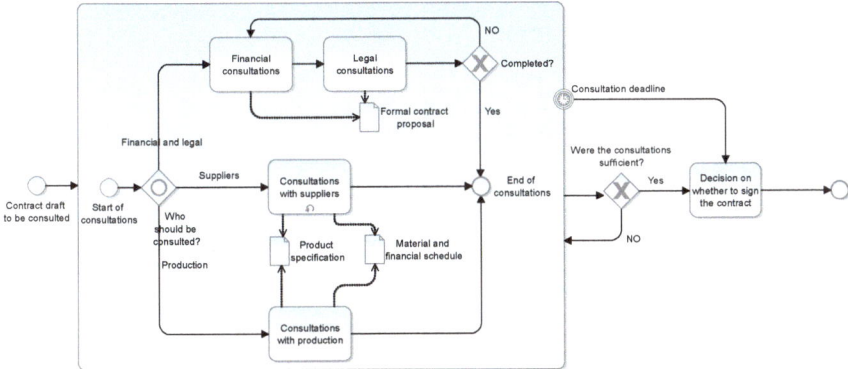

Fig. 4.6 Example of static business process modeling which takes into account the context of events (process: Consultations prior to signing the contract). *Source* Author's own elaboration

and this increase—unfortunately—will result in wrong decisions being made (and their tangible negative results). Of course, each of the scenarios is just optimal "as of today." As the previous example shows, however, with time all of them will accumulate the risks and drawbacks of becoming less clear.

The BPMN standard allows to extract the specific types/contexts from the main decision-making process in the form of sub-processes, which are triggered by specific events.

As before, such a solution requires to define all of the scenarios/contexts of signing an agreement in the course of the modeling stage (in the example on Fig. 4.7: all of the contexts of signing the contract). However, this solution still runs the risk of being incomplete or too complex. However, it leaves us with a more transparent graphical diagram and enables us to make use of predefined sub-process templates with ease. In practice, another benefit during implementation is that the sequence of executing subsequent sub-processes depends on the time of their appearance and the priorities assigned to each sub-process in the BPMS system.

4.4.2 Risks Associated with the Static Modeling of Processes Which Are Dynamic in Nature

All of the above examples run similar risks, which are characteristic of the static modeling of processes which are, in fact, dynamic in nature. Such attempts usually lead to implementation projects running over time (rising costs, more additional risks) and falling employee morale (no sense of responsibility for one's own work), and may have the unintended effect of lowering efficiency within the organization [19]. Certainly whenever processes are associated with clients and the market in particular, it is crucial to establish whether dividing the process in detail into individual

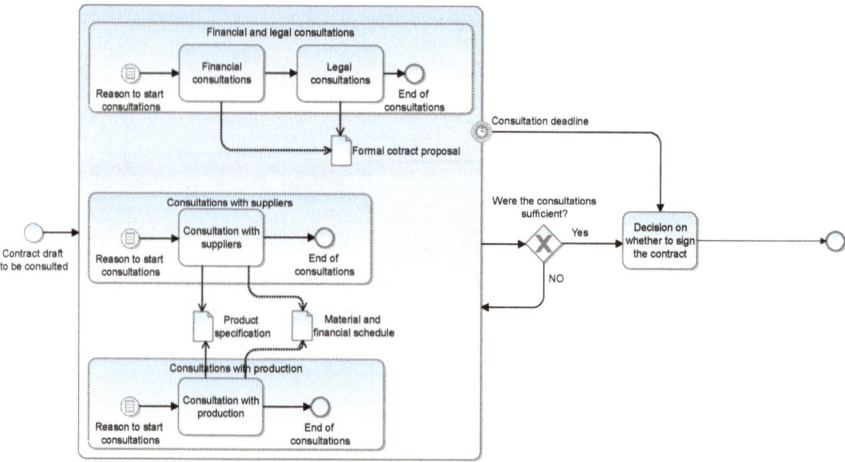

Fig. 4.7 Example of static modeling of a decision process with the use of event sub-processes (Process: Consultations prior to signing the contract). *Source* Author's own elaboration

"indivisible" tasks, which can be performed by their direct performers, will not lead to higher losses due to the over-complication and over-specification of processes and, first and foremost, due to preventing employees from acting in unforeseen circumstances—even when the employees in question would know how to perform the work at hand.

The main risks associated with such an approach are:

- **losing the transparent and flexible character of processes as the result of their over-specification**

This results in the "creeping" over-complication of processes due to adding different special exceptions, "contingency plans," and conditions which "should" be taken into account despite the fact that they only arise in special circumstances.

In the nineties, the author worked on a settlement system for foreign contracts for one of the international trade centers. At the express request of the project leader, he has taken into account all of the possible scenarios of performing a contract. The program even factored in the option of performing a contract in two countries at the same time, in ten different places, and using three different currencies, only because one such contract had been performed in the span of the last 25 years! However, in practice it turned out that even such a system did not take into account all of the possible circumstances, as one situation arose which was not factored in the system at all.

- **strengthening a culture of unaccountability**

The strict imposition of an unchanging method of process execution, which does not factor in changing circumstances, rids employees of their initiative and takes away their accountability for the results of the process. Not only that, it even encourages them to accept a situation which causes losses—but which follows the statute/procedure/process! After all, if the statute/procedure/process/owner is responsible for the outcome, the employees think it wise to stay close to the process, even in the knowledge that such actions are senseless, suboptimal, and will result in losses.

- **the systemic inclusion of unaccountability in the implementation of process management**

Detailed static process modeling may be used as a chance to create a system, in which formal accountability is moved upwards to higher levels of management, despite the fact that it will not cause the practical reduction in the scope of the privileges and competences. This pertains to both the privileges of the employees, as well as the privileges of entire organizational units under their direction: it is enough to include in a large number of processes that the individual responsible for the unit is also responsible for the decisions pertaining thereto (CEO, manager, etc.). It can be said with a high degree of certainty that preparations for the decisions will still remain in the hands of the employees, but it will be their superiors who will bear systemic accountability for their results.

> In one of the companies implementing the Purchase management process—at the express request of the financial manager and with the blessing of the employees themselves—the prepared process model designated the financial director as the sole individual responsible for all purchase decisions. After the launch of the process-driven application supporting process execution, on the first day the financial manager has received over 60, and on the second—over 100 individual decisions to make. Of course, even if he were to entirely abandon his other duties, he would still be unable to thoroughly understand the rationale behind each decision that he was expected to "make" and for which he was to be accountable.
>
> The process models have been changed on the second day of operations.

- **active resistance against implementation**

The introduction of static processes as decision-making algorithms results in the further detachment of the place of making decisions from the place of actual work performance. For some of the knowledge workers, cognizant of the character of their work, the common-sense will to keep privileges pertaining to the decisions on the best method of performing their work will lead to perceiving change as a threat and—inevitably—result in their active resistance thereto.

In the course of building clinical pathways and implementing process management in healthcare units, the author has confirmed time and again that using the word "algorithm" in conversations with medical doctors, as well as his reluctance to decisively and unambiguously refute suspicions and accusations of his intention to "implement algorithms," will immediately result in quick, coordinated actions on the part of the medical doctors in order to block the implementation, or even result in the immediate termination of the mutual project.

As has been demonstrated, methods of describing dynamically managed processes which are at odds with the requirements of the users may, in the course of implementation, lead not to the revealing and creation of knowledge, but to its loss; and not to the strengthening of the culture of accountability, but to its weakening. For this reason, it is essential to ask ourselves: *Is it possible to describe (model) dynamic business processes, and if so—how?*

4.4.3   Dynamic Business Process Modeling

BPMN 2.0.2 offers the possibility of modeling a dynamic subprocesses which allows us to perform multiple non-structured and non-prioritized, but predefined actions [20]. Within an ad hoc sub-process, such actions can take the form of:

- tasks, the equivalent of a checklist of available actions

  or

- short sub-processes, the equivalent of a set of predictable business scenarios, usually containing predefined, reusable process patterns.

The actions can involve the performer of the ad hoc sub-process, as well as other individuals (or process roles). In the case most implementations of BPMS systems (e.g. Bizflow, Fujitsu, IBM), the process performer is also authorized to perform actions which were not taken into account in the initial process modeling stage [21]. As in the case of each action, here as well performers of ad hoc processes are authorized to make use of documents, data objects, or virtually any other task artifacts or attributes which are included in the standard or defined as needed in accordance with BPMN notation.

The diagram from Fig. 4.8 enables the process performer to perform selected consultations (one consultation, a selected number of consultations, or all of them) on a one-off basis or on a repeated basis as required. When we define:

"||"—as consultations held at the same time, and
"→"—as consultations held in sequence,

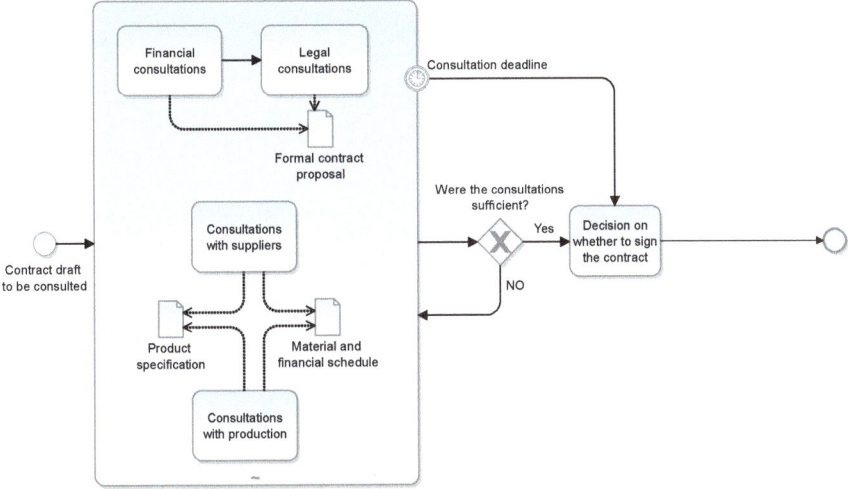

Fig. 4.8 Example of dynamic decision-making process modeling (Process: Consultations prior to signing the contract). *Source* Author's own elaboration

we can describe sample scenarios of performing consultations within the process described in the diagram from Fig. 4.8 as follows:

finance* → production → supplier → finance* → production
or
(supplier ‖ production) → finance* → production → (finance* → supplier) ‖ production

In the modeling stage, it was assumed that each financial consultation will be followed by a legal consultation as a matter of obligation (finance* = finance → legal). Of course, each of the tasks can be included as a separate item or in any combination thereof in accordance with the actual course of the business process or requirements resulting from e.g. risk analysis.

As can be concluded from the above example, a model of a dynamic process is a set of tasks which can be performed. It is possible—though optional—to define relations between selected tasks. The specific course of the process is decided by its performers—the knowledge workers. Only the documentation of the completed specific performance shows the process as a specific sequence of events. Acknowledging this fact should lead to the extension of the process description notation (e.g. version 3.0 of BPMN) with the possibility of defining processes or their unstructured parts in the form of a multilevel checklist assigned to a specific process role or group of roles. This solution will be:

• closer to the experience accumulated over multiple years in e.g. the aviation industry, the healthcare industry, or the construction industry;
• much more ergonomic in the process discovery and communication stage;

state	task	deadline	product
○	Financial consultations Legal consultations	Jan 3rd, 2018 Jan 3rd, 2018	 Contract proposal
○	Consultations with suppliers	Jan 10th, 2018	Specification of the component Supply schedule
○	Consultations with production	Jan 15th, 2018	Specification of the product Production schedule

Fig. 4.9 Example of a dynamic model of the decision-making processes in the form of a simple checklist. *Source* Author's own elaboration

- closer to the implementation of business process support in MRPII, ERP, CRM, or HIS/EMR systems, in which the user interface is more likely to contain checklists than process diagrams (e.g. in EPC, UML, or BPMN notation).

A sample notation in the form of checklist for the decision-making process checklist from Fig. 4.8 is shown on Fig. 4.9.

From the perspective of the user, the postulated expansion of the BPMN notation would also allow for the much closer integration of dynamic BPM with case management. Because at present both of these methodologies already enable not only the use of static "algorithms" to describe the course of processes, but also the modeling and execution of processes which are dynamic in nature, designing a single, shared notation for both would enable us to prepare uniform process descriptions and process-driven applications encompassing a broad spectrum of processes within the organization. This would allow organizations to raise the transparency of process description with a wide range of knowledge workers in mind, eliminating the requirement of learning two (or more) different notations, as well as having to decide which notation to use in which tool and for which group of processes.

As has been discussed in the section Searching for the possibility of adapting traditional process management to the requirements of the knowledge economy in Chap. 1, at present most BPMS systems already offer the possibility of describing and modeling dynamic processes, and CMS systems offer the possibility of modeling static processes. This prevents us from having to own and use two different process systems at once in the organization, and allows users to work within a single, uniform application work place. Furthermore, it allows us to analyze completed processes and to learn from them regardless of the type of processes involved [22].

Instead of complicating and expanding the Contract process with additional consultations and their logic conditions and priorities, the possible options were described in the form of a much clearer diagram (Fig. 4.8) or checklist (Fig. 4.9), empowering the process performer to perform the process in accordance with the requirements of a specific performance. Such a solution supports natural, organic work, which draws from the engagement, creativity, and knowledge of the employees. Furthermore, it reduces the risks associated with performance thanks to the full transparency and ongoing oversight of ad hoc actions, preventing the aforementioned "hidden factory" effect.

As we can see from the above example, the fundamental aim of the modeling process has changed. It departs from the traditional aim of defining and optimizing the specific sequence of actions as a one-off process. Instead, in the case of dynamic BPM the aim of the modeling process is to define a set of tasks, and in the case of predictable processes or predictable parts of processes—to define the standard sequence in accordance with the best knowledge of the organization, as well as to define the method of performing the process in non-standard conditions. In the knowledge economy, another goal is to accumulate knowledge on the full context of a specific performance, which is crucial in the analysis stage. We understand the term "method of performing a process in non-standard conditions" as defining:

- privileges to choose tasks specified in ad hoc sub-processes (tasks should not be available to all individuals),
- privileges to perform tasks not specified in ad hoc sub-processes (the freedom to deviate from the standard process, that is, perform limited experiments, should not be available to all individuals),
- limitations of particular tasks (e.g. parallel/serial performance of ad hoc tasks, time or resource limits (a process cannot have an indefinite time and engage all possible resources).

Though the above requirements were formulated as the result of analyses of the performance of dynamic processes, they are practically identical with the requirements and principles defined in adaptive case management, which have been described in Chap. 1.

4.5 Simulations of and Optimization Research on Dynamic Processes

Similar to traditional business process management, in the case of dynamic business process management one of the elements of the Process (re)Design stage consists of exploratory and simulation research, which allow for the verification in controlled conditions of the consistency between the prepared process description, the initial conditions, and specific past performances of the process. The aim of this section is to seek answers to the following questions:

- *Is it possible to perform simulation studies on dynamically managed business processes, including unpredictable processes?*
- *What are the benefits of such studies?*

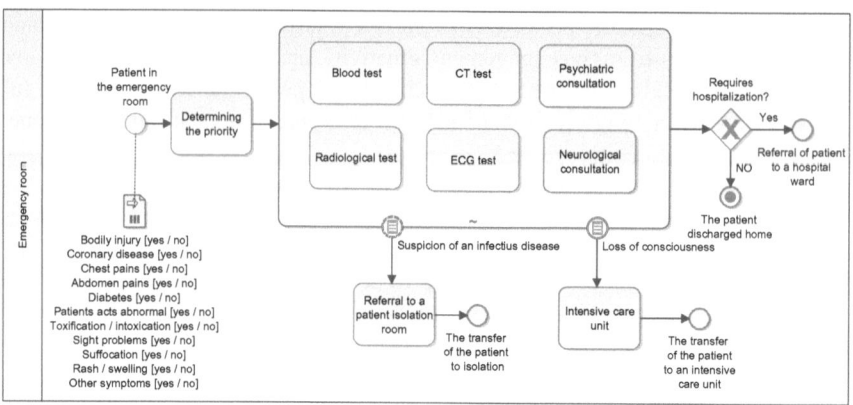

Fig. 4.10 Dynamic process modeled in the BPMN notation

It has been assumed to date that simulations may only be performed in relation to static processes, as they typically refer to previously identified process pathways, which are dependent on the assumed probabilities of making certain decisions or the occurrence of certain events which might influence the course of the process. In other words: the modelers of statistically modeled process defined parameters for the performance of specific tasks and probabilities of making known and described decisions, and then went on to evaluate and optimize the results of process execution (e.g. average time of performance, used resources, or production costs). Such an approach is only partly useful in the analysis of dynamic processes, which is interested not only in average times or costs, but also (and sometimes even first and foremost) changes made to the course of a process, which are the result of a specific, individual performance. In effect, we are not only interested in the probabilities of specific performances, but also in the methods of performance and their influence on the results of performing a dynamically changing process. The aim of such analysis is to build best practices pertaining to process execution, as well as to point to the risks resulting from each method of performance.

To this end, we can use tools which enable us to model and simulate the course of the process in accordance with criteria pertaining to data which influences the performance. Unfortunately, simulations cannot directly make use of dynamic processes, e.g. processes modeled with the use of ad hoc constructions in BPMN. It is crucial to define the method of simulating the principles that an individual in the role of a process performer would draw on. In the case of the process model presented in Fig. 4.10, it was assumed that it is the process performer who decides on the specific method of performing the ad hoc process, which consists of the following tasks: Blood test, CT scan, Psychiatric consultation, etc. In the course of the simulation, the performer's choices are replaced by:

- a group of functions pertaining to data processing

  and

- a group of functions controlling the course of the process in accordance with changing data values.

For this reason, in the course of preparing the simulation it is essential to supplement the process model with input data structures, which will determine process execution (either directly or after being processed in accordance with the aforementioned functions) (Fig. 4.11).

The preparation of a dynamic process model to undergo simulation requires:

- defining the set of input data (often along with statuses), which will be available in each instance of process execution;
- defining the resources available for process execution (often in the context of the schedule);
- defining criteria of changing the status of data while performing particular steps;
- for tasks performed in loops—defining the criterion of exiting the loop;
- allocating the time of performing each action (sometimes along with formulas making the allocated time dependent on other data);
- defining the formulas (functions) indicating the pathways selected during process execution. Sometimes costs of resources and actions (if the study pertains to e.g. costs of process execution) and other parameters, e.g. criteria of success/failure, are also defined;
- defining principles (privileges) and the frequency of the need to perform unpredictable tasks which are not contained in the process diagram (e.g. sudden mass incidents, equipment failure, or medical personnel becoming indisposed);
- defining the frequency and other criteria of the appearance of tokens (e.g. the number of patient during a medical shift).

Simulation research offers us the possibility of analyzing the possible sequences of performing tasks in the course of a given process, as well as the possible consequences of their performance. The formulas accepted in the simulation (e.g. probability distributions for events or decisions) can be verified on the basis of information about past performances, which has been collected as the result of exploring data from transactional systems, e.g. domain-specific systems or integrated ERP or HIS/EMR systems. In effect, it is possible to calibrate process simulation models on the basis of data from past performances. However, one should note that process discovery techniques enable us to perform analyses regardless of how processes were performed to date. In the course of performance, specific tasks are logged in an event log of the owned system (BPMS or CMS), and their results—entered in relevant data objects. By reading the time stamps for specific performances included in the logs of the BPMS/CMS system, we are able to accurately determine who performed which tasks and in what sequence—and what were their results. In effect, we are just a step from building a knowledge database on subsequent performances, their courses, used or collected data, results, and the performers who in the course of performance gained specific

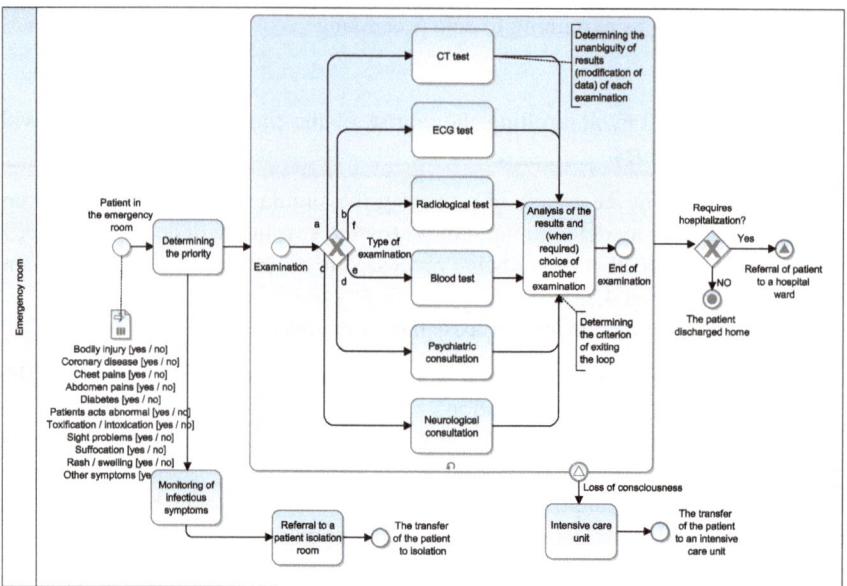

Fig. 4.11 Dynamic process prepared for simulation

knowledge (information, data on the context of performance, and experience). This shows how processes were performed under the conditions of performance to date.

However, what is often of more interest in the knowledge economy is the issue of how to perform processes in conditions that we believe are probable (e.g. as the result of planned organizational changes or changes to production) or certain (e.g. at the result of changes in the law). Simulations allow us to establish how dynamic processes can be performed in new conditions, what are their potential results, and what is their probability. In effect, we are able to analyze what can happen in the course of a dynamic process before it happens, and which process parameters lead to optimal results. Due to the fact that dynamic processes are often dependent on multiple parameters which independently influence one another, without simulations it would be hard to determine which method of performance will lead to the most beneficial results. Multiple comparisons of the results of simulations with later results of data exploration pertaining to actual performances allows for the increasingly more accurate calibration and more down-to-earth design of models used in the simulations. In effect, despite the unpredictable nature of dynamic processes, it is possible to build databases of best and wrong practices on the basis of not only actual performances, but also predicted changes in the conditions of performance or predicted contingency scenarios.

4.6 Different Forms of Process Descriptions in Different Stages of the Process Lifecycle Within the Organization

Process descriptions and diagrams created in the course of the Process discovery and rationalization stage in the further part of the implementation process and process management have the function of a repository of information used and updated on an ongoing basis. As has been demonstrate in Sect. 4.2 of this chapter, in the case of traditional process management the fundamental decision rests in the choice of the form and the notation in which to describe business processes [10]. In doing so—in the case of dynamic BPM in particular—one must remain cognizant of the following contexts arising in connection with the entire process lifecycle within the organization:

- the preparation and dissemination of process documentation;
- preparation and testing of process-driven applications;
- the use of business process descriptions in the course of process execution;
- the evaluation of the results of business process execution;
- the innovation of processes and the ongoing actualization of their documentation.

Process documentation should allow for the transparent presentation of the knowledge contained in process descriptions, but also support process execution itself, as well as ongoing process evaluation and the ongoing verification and updating of knowledge. In accordance with the views of Pucher [16, pp. 68–79], process documentation should include:

- process goals;
- process data;
- subject matter mentioned by participants of the process;
- roles (actors) included in process execution;
- a description of interactions between participants of the process.

Due to the unpredictable nature of process execution, the documentation of a given process must allow for the introduction of changes to each of the elements of the process in the course of implementation, performance, or *ex post* analysis. The documentation should also account for the practical conditions of performance, and in effect—different requirements in regard to access to information and different conditions of introducing changes to process descriptions. Given all these changes to the conditions and requirement of a dynamic process, should it have the same form of description throughout the entire process lifecycle? Why should the form of description remain constant when the needs of the process performers are changing? The aim of the studies was to confirm that changing the form of process description in the course of the process lifecycle is not only possible, but also expected by the process performers themselves.

The first argument in support of introducing the possibility to change the form of description lies in the fact that in the past (as in the present), management resorted to multiple methods of describing business processes, sometimes ones which did not even refer to the term "process" at all. At present, the most common forms of process descriptions are [23–27]

- textual description;
- structured description (e.g. in the form of a procedure);
- flowchart;
- block diagram;
- table;
- control list (checklist);
- process diagram;
- Gantt chart;
- 3D process map.

As the description of a diagnostic-therapeutic process from Fig. 4.12 demonstrates, it is also possible to combine several of the abovementioned forms within a single description.

In effect, there exists a long-known need for different forms of process description. The selection of a specific form of description should not be limited to notations describing diagrams of business processes, such as BPMN, BPMS, or EPC [29].

In the course of a 2012 project on business process identification in a large enterprise from the Polish construction sector, the author, following a training course on business process modelling in accordance with BPMN, advised the Quality Management Officer to change the form of describing business processes in the quality documentation from flowcharts to BPMN diagrams. In response, the Officer said that the former method of description has been used in the last 30 years and hence flowcharts are understood by the entire ten-thousand-people-strong employee base of the enterprise. The Officer made the practical assessment that changing the notation to BPMN would be beneficial only for select groups of processes and the implementation would encompass a year in the least, if not two—not because of the time needed to implement the modeler and train the process modelers themselves, but because of the time needed for the entire employee base to understand the new models. In effect, the narrower implementation has been successfully performed.

The above example clearly demonstrates who the actual recipients of process models and descriptions are, as well as points to the value of the pragmatic aspect of process modelling for further ongoing practical use of the processes in question [30, 31]. It is crucial to accommodate the modelling language to the field in question and the level and degree of competence of the recipients and performers of the models [32]. Similar to corporate architecture, in multiple instances specific groups of performers should be provided with different presentations of process models and descriptions, which for the sake of clarity and transparency should only concern crucial information or a form of process description tailored to those specific groups

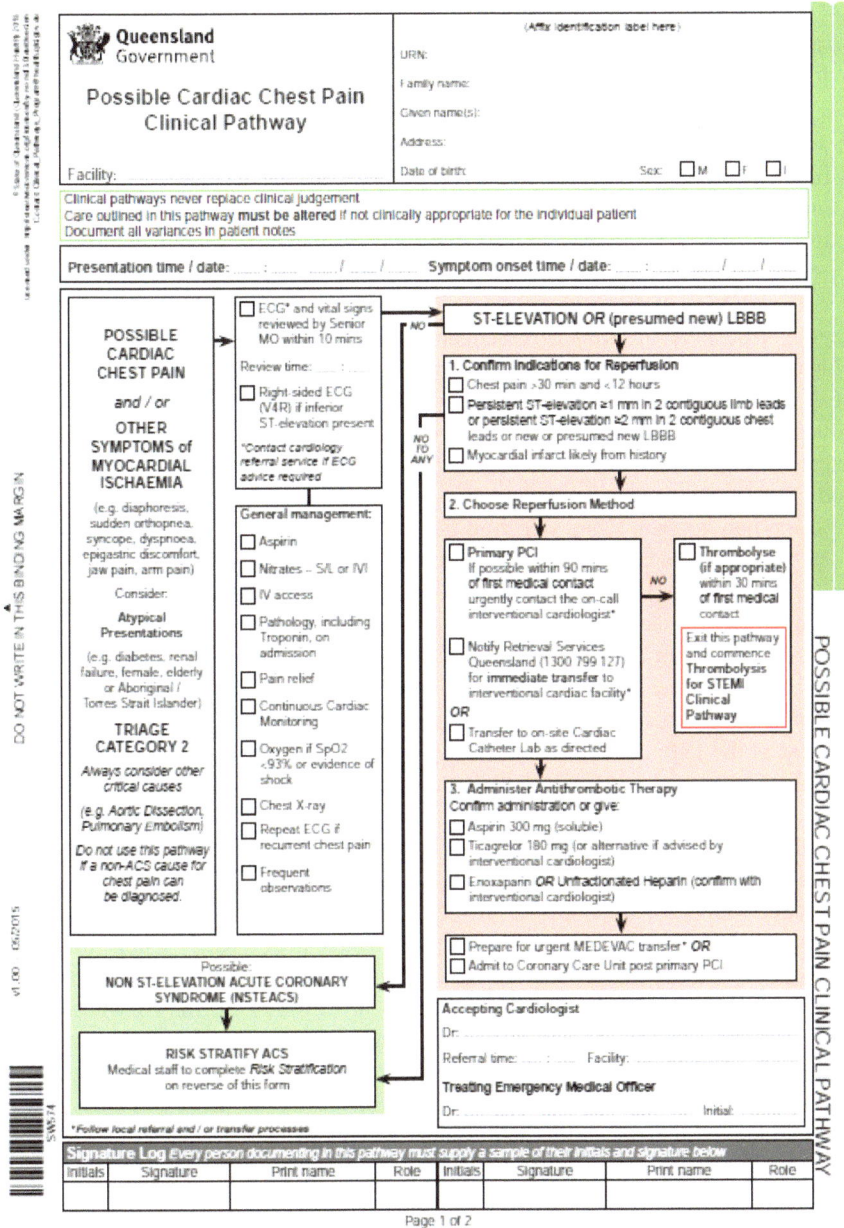

Fig. 4.12 Example of a diagnostic-therapeutic process description. *Source* State of Queensland [28]

of recipients [33]. After all, process descriptions and models are not prepared with their modelers in mind, but rather, with a view of a wider scope of practitioners, who will use them in their ongoing work and who will, on the basis of their experiences, propose innovations to the processes in question. In effect, it would be irresponsible to select the method of the form of notation for process descriptions in light of the beliefs and habits of a select group of "process specialists" alone. Such a choice would run the risk of the knowledge repository on processes becoming useless due to the non-transparent nature of the information to the wider employee base.

As has been demonstrated in Chap. 1, as recently as six years ago it was believed that the modeling of dynamic processes is outright impossible [17]. To a considerable degree, especially in terms of practical implementation, creators of new methodologies and IT solutions had to overcome the problem of forms of descriptions of dynamically managed business processes. In the case of traditional, static processes the issue is considerably less important. In their case, the process performer is expected to perform the process in accordance with a prepared model ("algorithm") and is prevented from introducing changes thereto. The only factors which should be taken into account when selecting the form of description in such a scenario is its adequacy to the field in question and the transparency and comprehensibility of the descriptions for the performers themselves. In the case of dynamic business process management, selecting the form of process description should be predicated on the knowledge that the process performer is authorized to introduce changes into the process in the course of its performance itself, and that such changes must be documented in the course of performance as well [20, 34, 35]. The lack of market success of the Ultimate BPM Suite Version 7 modeler, which offered the option of changing the process diagram in the course of performance itself (advertised as the possibility to introduce changes to the process diagram "on the fly") demonstrates that the BPMN diagram as the form of process description, while effective in the process identification stage, might nonetheless prove to be ineffective during performance itself [20, 34, 36, 37]. At the same time, the practical success of tools built around methodologies of adaptive case management (ACM) demonstrates that knowledge workers performing dynamic business processes prefer other forms of process description than traditional diagrams in BPMS, EPC, or BPMN notation [38]. In effect, attempts at forcing the performers to use notations considered to be "genuinely process-based" in nature (as they are based on process diagrams) are doomed to fail, especially in light of the character of the processes (the field of modelling), but also due to the performers' own habits.

According to Hammer, organizations which wish to be process driven need to discover (identify), communicate, measure, and manage processes [39]. As has been demonstrated in Chap. 2, the business process lifecycle is usually divided into the following four main stages (Figs. 2.1, 2.2 and 2.4), such as:

1. Process (re)Design (discovery and modeling).
2. Implementation and adjustment.
3. Performance and monitoring.
4. Analysis and diagnosis (process improvement).

Process descriptions are being created, used, and modified in each stage of the lifecycle, albeit usually in different conditions and by different employees within the organization. The Process discovery and communication stage is usually not time sensitive and is dependent on a select team of trained individuals. In the case of dynamic processes, it is not practicable to describe all of the possible paths a process could take in detail, nor is it practicable to describe all of the factors which might influence its performance. As has been described in Sect. 4.3 of this chapter, such an attempt at including all of the aforementioned details could significantly impair the description's comprehensibility, and thus significantly lower its practical value.

In the Process implementation and adjustment stage, process descriptions must first and foremost be clear to the employees, who have different levels of knowledge, including different levels of experience in terms of process execution. In effect, they must also be understandable to newly hired employees or employees working in organizational units which do not directly participate in process execution (e.g. accounting or controlling), but which should nonetheless understand the principles behind their execution. Processes should also be capable of being used as a means of familiarizing external partners (e.g. suppliers or subcontractors) with their roles as participants in processes or as individuals whose work influences the processes in question. In this stage, work with process descriptions is usually not time constrained, but different groups of recipients may have different expectations in regard to the transparency, scope of information, and level of detail of the presented process descriptions.

In the Process execution and monitoring stage, process descriptions are used by a broad range of employees within the organization. In effect, their form should be tailored to the employees' habits and skills (the pragmatic aspect of accommodation). In the case of dynamic BPM, it is essential to account for the possibility to adapt or even individualize processes in the course of their performance. Such a solution should include the options to:

- omit certain tasks included in the process description;
- add new tasks or parameters which are not included in the process description;
- indiscriminately alter the sequence of the performed tasks;
- perform certain tasks repeatedly—both included in the process description and added by the performer.

In the Analysis and diagnosis stage (process innovation), traditional process management allows for the evaluation of key performance indicators (KPI) for individual tasks, their groups, or the entire process itself. In the case of dynamic BPM it is also essential to allow for the performance of a comparative evaluation of the process execution and its model, as well as a comparative evaluation of multiple performances of a given process, with the aim of not only establishing whether the performance went in accordance with the model, but also building upon and innovating on the model itself [40].

The aim of the study performed by the author was to establish which pragmatic aspect and which stage in the process lifecycle influences the expected form of description of business processes within the framework of dynamic business process management over the entire process lifecycle within the organization. The study answered the following question: *"Should a dynamic business process have the same form of description throughout the entire process lifecycle?"* and decidedly falsified the thesis that *"The form description of a business process should remain constant throughout the entire process lifecycle."*

4.6.1   The Research Methodology

A diagnostic-therapeutic process has been selected as the subject of the study, as it is undoubtedly a dynamically managed process; one which requires each subsequent performance to be tailored to the specific context thereof, and one whose each performance verifies current knowledge and creates new knowledge. It is also a process of considerable practical significance.

The research had the form of two independently held surveys in the multi-specialist St. Padre Pio Provincial Hospital in Przemyśl. The first part of the research was performed with the participation of medical doctors from the Pediatrics Ward, while in the second—the medical management staff (Heads of Hospital Wards). Prior to administering the surveys, with the participation of each supervisor, that is, the Head of the Pediatrics Ward and the Deputy Health Care Director, a short meeting was held in each department in order to explain the aim of the research and all of the terms used in the study in the context of the daily work of a medical doctor. Furthermore, the surveys contained Keys with definitions and providing examples of the use of all of the terms used therein.

In the first stage of the study, the participants were asked about their preferred form of description of the diagnostic-therapeutic process during:

I.   patient diagnosis and treatment planning;
II. patient treatment, including potential modifications of the individual treatment plan (ITP);
III. *ex-post* evaluation of performed diagnostic-therapeutic processes.

For each stage of the process lifecycle, all eight forms introduced in the first part of the article were suggested as the possible forms of process description (verbal description, structured description, flowchart, table, control list, process diagram, Gantt diagram, and 3D process map).

In the second part of the study the participants were asked about the key determinants behind their selection of the form of description of the diagnostic-therapeutic process in the course of its entire lifecycle, which, like in the second part, was divided into stages (diagnosis and planning, treatment, *ex-post* evaluation).

4.6.2 Results of the Study

The study has demonstrated that regardless of the method of identifying the diagnostic-therapeutic process, in the course of patient treatment planning the form of presentation preferred by medical doctors of the ward in question and the Heads themselves was not a process diagram, but a checklist (43% and 31%, respectively). For the Heads of wards, the second-best preferred form of description was a block diagram and process diagram (18% each), and for the medical doctors from the ward in question—a structured description (15%).

Even when the description of an individual treatment plan was prepared in the form of a process diagram, in the course of patient treatment itself over 40% of medical doctors who took part in the survey prefer working with a process description in the form of a checklist. 50% of the respondents from the ward in question (though only 26% Heads) find it easier to modify a stage III clinical pathway representing an individual treatment plan in the form of a checklist in comparison with a process diagram, or even in comparison with a simplified flowchart. In this stage of the process lifecycle only 5% of doctors consider working with process descriptions in the form a Gantt diagram.

In stage III (ex-post evaluation and modification of the performed process) only 18% of the medical doctors from the ward in question and only 27% of the ward Heads selected a process diagram. The majority of the medical doctors (almost 31%) selected the Gantt diagram, which was not considered in the previous stage, as their preferred method of presenting the diagnostic-therapeutic process. The Gantt diagram facilitates the evaluation of the resources used, the relations (connections), as well as the timeline of subsequent performed tasks. When information on the duration and number of tasks is included in the description, the Gantt diagram allows for easier visual reference of the tasks at hand (Table 4.3).

In the second part of the study both the medical doctors from the ward in question and the ward Heads indicated which determinants influenced the choice of their preferred form of process description. In all of the stages of the diagnostic-therapeutic process, for the medical doctors the main determinant (or one of the main determinants) of their choice was "being able to evaluate the treatment on an ongoing basis" (a total of 27%). In the Patient diagnosis and Treatment planning stages, the determinants "being able to evaluate the treatment on an ongoing basis" and "selecting the clearer form of presentation for patient data in order to facilitate making the right clinical decisions" have been given equal value. For the Heads of wards, in the Treatment planning stage, the most important determinant was limiting errors by way of evaluating the treatment on an ongoing basis (27%). In the Patient treatment and Ex-post evaluation stages, the main determinant was the possibility to easily compare the treatment with anonymized patient data from other treatments (20% and 30%, respectively).

In total, the study determined the following factors to be the main determinants of the form of description of diagnostic-therapeutic processes throughout their entire lifecycle:

Table 4.3 Forms of description of the diagnostic-therapeutic process preferred by medical doctors

Stage of the lifecycle of the diagnostic-therapeutic process	Preliminary diagnosis and patient treatment planning	Patient treatment	Ex post evaluation of executed diagnostic-therapeutic processes
The main aim of using a process description	Using available knowledge (including experience) to prepare an effective treatment plan The patient and the closest relatives understand the treatment plan	Effective treatment of the patient with the use of all available resources and knowledge	Evaluation of all potential risks and planning of further treatment (e.g. outside the healthcare unit) Verification of old knowledge and identification of new knowledge
The main features of the context of process execution	Usually no direct time pressure Being able to consult and modify treatment plans a number of times	Work under the pressure (or extreme pressure) of time constraints The need to accommodate the initial plan to the progress of a specific treatment, including unforeseeable circumstances Responsibility (most decisions are irreversible)	Lack of time constraints Being able to consult, modify, or supplement the results of evaluation a number of times
The form of description for the diagnostic-therapeutic process preferred by the medical doctors	1. Checklist	1. Checklist	1. Gantt diagram
	2. Process diagram	2. Process diagram	2. Process diagram or Table
	3. Structured description or Flowchart	3. Flowchart	3. Flowchart

Source Author's own elaboration

- Being able to evaluate the treatment on an ongoing basis (21%),
- Being able to dynamically modify both planned and ongoing tasks (20%),
- Limiting errors through evaluating the treatment on an ongoing basis (17%) (Table 4.4).

Table 4.4 Determinants of selecting the form of description for the diagnostic-therapeutic process as indicated by medical doctors

Stage of the lifecycle of the diagnostic-therapeutic process	Preliminary diagnosis and patient treatment planning	Patient treatment	Ex post evaluation of executed diagnostic-therapeutic processes
The main aim of using a process description	Using available knowledge (including experience) to prepare an effective treatment plan. The patient and the closest relatives understand the treatment plan	Effective treatment of the patient with the use of all available resources and knowledge	Evaluation of all potential risks and planning of further treatment (e.g. outside the healthcare unit) Verification of old knowledge and identification of new knowledge
The main features of the context of process execution	Usually no direct time pressure. Being able to consult and modify treatment plans a number of times	Work under the pressure (or extreme pressure) of time constraints. The need to accommodate the initial plan to the progress of a specific treatment, including unforeseeable circumstances. Responsibility (most decisions are irreversible)	Lack of time constraints. Being able to consult, modify, or supplement the results of evaluation a number of times
The determinants of the selection of the form of description preferred by the medical doctors	1. The possibility of evaluating patient treatment on an ongoing basis	1. The possibility of evaluating patient treatment on an ongoing basis	1. The possibility of easy comparative analysis with anonymised data from other patients' therapies
	2. The possibility of introducing dynamic changes to planned and performed actions	2. The possibility of introducing dynamic changes to planned and performed actions	2. Reducing operational errors through monitoring
	3. Reducing operational errors through monitoring	3. Reducing operational errors through monitoring	3. The possibility of analyzing the used resources and their level of productivity

Source Author's own elaboration

4.6.3 Results and Discussion

The results of the study clearly demonstrate the variability of the medical doctors' expectations toward the form of description of the diagnostic-therapeutic process. It is no longer possible to assume that the description of a business process should remain unchanged throughout its entire lifecycle. This variability is the result of both different needs in regard to the process description at different stages of the process lifecycle, as well as the different contexts in which the process performers find themselves (time constrains, available data, the availability of HIS/EMR systems, etc.).

Actions in the Patient treatment stage, or the performance and ongoing evaluation of a dynamically performed process, are in most cases undertaken under time pressure. In such a case it is often impossible to evaluate process diagrams or process descriptions in table form, much less to read multiple-page procedures with structured descriptions. In such circumstances, there is even less time to update process diagrams. For this reason, in this stage most medical doctors prefer process descriptions in the form of a checklist, which facilitates the ongoing evaluation of treatment thanks to its transparent, intuitive form of presenting information. Checklists facilitate the making of correct clinical decisions and are a form which can be modified and extended with new tasks with relative ease [41] (Table 4.5).

Actions undertaken in the *ex-post* evaluation stage are usually performed without time constraints and with the patient in stable condition. Their primary aim is the identification of the potential threats which could influence the patient's further treatment. From a broader perspective, however, their goal is also the verification of the efficiency of using current knowledge and the identification of new knowledge, which stems from the performance of the diagnostic-therapeutic process. In light of this fact, the choices of the medical doctors participating in the study seem understandable. The description of processes in the form of a Gantt diagram supplemented with fundamental parameters detailing the patient's health enables the respondents to make a thorough evaluation of the timeline of the performed actions and their results within the framework of the treatment process of an individual patient. In turn, a 3D process map allows for the graphical comparative evaluation of a clinical pathway (the standard process), an ongoing or completed treatment process for an individual patient, and multiple different treatments, about which data is stored in a given department's knowledge bases or in Evidence Based Medicine (EBM) databases.

As the aforementioned study results, the author's own experiences with consulting projects, and similar experiences of other scholars [42, pp. 163–180] demonstrate, the belief that the method of presenting a dynamic process (e.g. a diagnostic-therapeutic process, a crisis management processes, and increasingly more often also sales or investment processes) remains constant throughout the entire process lifecycle in accordance with the needs of the performer is false. The research has demonstrated beyond doubt that in the case of the diagnostic-therapeutic process the user expects the form of description to change throughout the entire process lifecycle. When describing such processes, therefore, one should take into account the following factors in selecting the form of presentation:

- the level of description and the character of the processes (the field that is being modeled);
- the group of recipients;
- the stage of the process lifecycle.

Further developments in regard to the implementation of dynamic BPM should concentrate on enabling the transposition of the form of presentation and the use of process descriptions in accordance with the needs of the user at any given stage of the process lifecycle. At the same time, the developments should enable the flexible redefinition, as well as the ad hoc creation of different forms of presentation specific

Table 4.5 The variability of the main goals of using process descriptions throughout the entire lifecycle of the diagnostic-therapeutic process

Stage of the lifecycle of the diagnostic-therapeutic process		Main purpose of the process description	Main features of process execution	Preferred form of process description
I	Patient diagnosis and treatment planning	Using available knowledge (including experience) to prepare an effective treatment plan The patient understands the treatment plan	Most work performed under time constraints Being able to consult and modify treatment plans a number of times	1. Checklist 2. Process diagram 3. Structured description or Flowchart
II	Patient treatment	Effective treatment of the patient with the use of all available resources and knowledge	Work under the pressure (or extreme pressure) of time constraints The need to accommodate the initial plan to the progress of a specific treatment, including unforeseeable circumstances Most decisions are irreversible	1. Checklist 2. Process diagram 3. Flowchart
III	Ex-post evaluation of the diagnostic-therapeutic process	Evaluation of all potential risks and planning of further treatment (e.g. outside the healthcare unit) Verification of old knowledge and identification of new knowledge	Lack of time constraints Being able to consult, modify, or supplement the results of evaluation a number of times	1. Gantt diagram 2. Process diagram or Table 3. Flowchart

Source Author's own elaboration

to the requirements of the users [43, p. 94]. As proponents of case management point out, it is essential that "process-based" scholars and practitioners strongly reject or actualize old knowledge and open themselves up to new aims and contexts of management, as well as new possibilities created through constant development of IT systems supporting management [35, 42].

4.7 The Implementation of Dynamic Business Process Management

As has been demonstrated in Chap. 2, the consequences of dynamic BPM are as follows:

1. The actual empowerment of process performers;
2. The radical acceleration of adapting to requirements;
3. A system of constant knowledge creation and verification;
4. The systemic dissemination of verified knowledge;
5. The systemic learning of the organization on the basis of daily experiences (a learning organization).

However, in order to reap the aforementioned benefits, organizations must adapt and implement a strategy of management in accordance with the three principles of dynamic business process management presented in Chap. 2:

The 1st principle of dynamic BPM   Comprehensiveness and continuity
The 2nd principle of dynamic BPM   Process execution should guarantee evolutionary flexibility
The 3rd principle of dynamic BPM   Processes are considered completed only after having been documented

This adaptation also requires the use of a methodology of implementation which exceeds the confines of traditional business process management and takes into account the consequences of dynamic BPM presented in Chap. 2, as well as changing

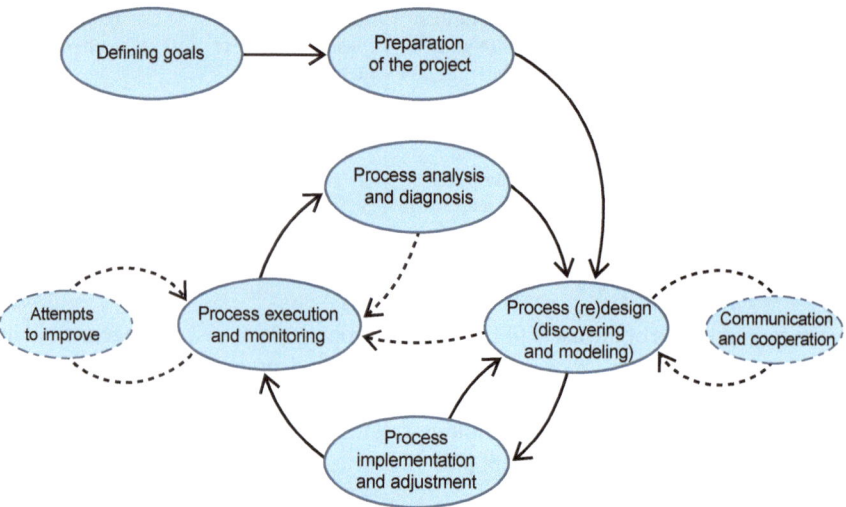

Fig. 4.13 The process lifecycle in accordance with dynamic business process management. *Source* Author's own elaboration

the approach to process discovery, a topic which has been discussed in Sects. 4.2–4.5 of this chapter. The basis of the methodology of implementing dynamic business process management formulated by the author lies in the process lifecycle presented in Chap. 2 (and repeated here for convenience in Fig. 4.13).

Necessary extensions to the paradigm of implementing traditional business process management are presented in Tables 4.6 and 4.7.

Table 4.6   The main actions undertaken in the course of preparing the implementation of process management in accordance with the concept of dynamic business process management

Stage	Goal	Actions undertaken in the course of implementing traditional business process management	Additional actions resulting from the concept of dynamic BPM
Defining goals	Agreeing on the goals of the implementation Minimization of the risk of misunderstanding the basic principles of process management	Defining and agreeing upon the goals of the implementation Defining the principles of implementation, including the level of engagement of the organization's top management	
Project preparation	The effective use of knowledge and tailoring the implementation to the process maturity of the organization	Overview of current internal (procedures, statutes) and external (e.g. legal requirements, licenses) rules and regulations. Defining the process maturity level of the organization Overview of IT resources, with a focus on maintaining ongoing support for business processes Preparation of a process description standard Preparation and approval of a process map Selection and preparation of training sessions for process owners and process managers and the implementation team Preparation and approval of a formal PID	*Overview of IT resources, including support for knowledge management* *Preparation of a map of essential knowledge resources available to the organization* *Analysis of the flow of knowledge within the organization, with particular focus on the sources of knowledge Preparation and approval of the principles of granting and communicating privileges to hold attempts at improving processes in the course of performance itself*

Source Author's own elaboration
Italic marks possibilities offered by dynamic BPM, but not available in traditional BPM

Table 4.7 The main actions undertaken in the course of implementing process management in accordance with the concept of dynamic business process management

Stage	Goal	Actions undertaken in the course of implementing traditional business process management	Additional actions resulting from the concept of dynamic BPM
(re)Design (discovery and modeling)	*Optimal method of implementing specific groups of processes*	Goals and goal indicators for processes should be defined prior to initiating process modeling. "Aimless" processes should not be modeled	*The nature of specific groups of processes and the resulting method of their description and modeling should be defined prior to process modeling*
	(For static processes and static processes with dynamic exceptions—development and approval of a standard process.	Preparation and approval of process models for selected processes in their current (*as is*) and improved (*to be*) versions in accordance with the organization's overall process model.	*Process descriptions should be designed (redesigned) on the basis of knowledge of the knowledge workers and data on past performances. Process descriptions should be combined or supplemented with descriptions of the flow of knowledge, with a particular focus on the sources of*
	For dynamic processes and static processes with dynamic exceptions—definition of principles and description of processes)	In the modeling stage, processes are usually rationalized, understood as the elimination of errors which were not identified prior to process modeling even by employees with years' worth of experience, and which are now obvious to all in the modeling stage Process diagrams should be unambiguous and understandable not just to process modelers, but first and foremost to all potential users, who will be obligated to make use of the information contained in process models	*knowledge, including "personal" sources of tacit knowledge* *Process descriptions should be designed in such a way, as to allow for varying forms of their presentations with different groups of employees in mind and in different stages of the process lifecycle*

(continued)

Table 4.7 (continued)

Stage	Goal	Actions undertaken in the course of implementing traditional business process management	Additional actions resulting from the concept of dynamic BPM
Communication and cooperation	*Broader use of intellectual capital thanks to the day-to-day sharing of knowledge* *Verification and creation of explicit knowledge on the basis of tacit knowledge in dialog with managers and experts* Dissemination of knowledge contained in process models and descriptions	The publication of process models and descriptions for all authorized people of interest, with a focus on the employees, but in justified cases also subcontractors, suppliers, or even the clients themselves The possibility of holding two-way communication with managers and experts responsible for particular processes with a view to: – verifying explicit knowledge available in the form of process models and descriptions (taking in questions and proposals for improvement), – using the tacit knowledge of the managers and experts to a broad degree	*The creation and ongoing access to a repository of knowledge within the organization, which includes process descriptions* *The creation of a community of practitioners (experts) engaged in designing and improving processes* *The possibility of using and objectively evaluating the knowledge and commitment of employees and experts on the basis of their actual engagement and results*

(continued)

Table 4.7 (continued)

Stage	Goal	Actions undertaken in the course of implementing traditional business process management	Additional actions resulting from the concept of dynamic BPM
Implementation and adjustment	*The aim is to prepare the organization to allow for the broadest possible use of intellectual capital thanks to the empowering and constant motivation of employees and the support of IT systems tailored to the needs of the knowledge workers*	Iterative use of process-driven applications, that is, applications in which the logic of operation is subservient to process diagrams. Changes to process diagrams and the creation of new process scenarios result in a change in the application's operation	*The preparation and approval of a prototype and a process-driven application on the basis of process description and principles of integration with existing ICT architecture*
			In the stage of creating prototypes of the user interface and an unintegrated application, both process descriptions and the user interface itself should be tailored to the requirements of the knowledge workers
	Raising the efficiency of operations with the support of ICT tools		*Preparation of mechanisms of innovation discovery (in the form of limited experiments) used by knowledge workers an the presentation of performed processes as a sequence of tasks (e.g. in the form of a process diagram or a Gantt diagram)*
			The quick, systemic identification of innovations and collection of knowledge

(continued)

Table 4.7 (continued)

Stage	Goal	Actions undertaken in the course of implementing traditional business process management	Additional actions resulting from the concept of dynamic BPM
Execution and monitoring	*The aim is the efficient performance of work and the creative use of the knowledge and dynamism of the employees in order to create and offer innovative products, services, and process improvements*	Ongoing monitoring and management of automated processes (standard control mechanisms). Monitoring and correcting the execution of the implementation plan, e.g. by the ongoing analysis of priorities and the selection of subsequent groups of processes to undergo Discovery and Automation.	*Performance of business processes with the support of process-driven applications, with the possibility of improving and creating processes in accordance with the requirements of a specific ongoing process and the privileges of the performers* *Ongoing monitoring and management of automated processes, including the analysis of the introduces improvements and their results. The possibility to analyze the entire context of the introduced improvements* *The verification and creation of knowledge. Quick, systemic identification of innovations, the creation of new knowledge, the rejection of old knowledge, and the dissemination of knowledge throughout the organization*

(continued)

Table 4.7 (continued)

Stage	Goal	Actions undertaken in the course of implementing traditional business process management	Additional actions resulting from the concept of dynamic BPM
Analysis and diagnosis	The dynamic management of business processes, not only their results	Ongoing controlling and management of automated processes Monitoring and correcting the execution of the implementation plan, e.g. by the ongoing analysis of priorities and the selection of subsequent groups of processes to undergo discovery, communication, and automation	*The ongoing and constant:* *(1) Discovery of the introduced improvements and the actual course of executed processes* *(2) Analysis of the results of the introduced improvements and executed processes* *(3) Quick dissemination of information on new or updated knowledge to authorized employees within the organization, with a particular focus on the direct process performers who might be influenced by the knowledge in question*

Source Author's own elaboration
Italic marks possibilities offered by dynamic BPM, but not available in traditional BPM

We should stress once more the fundamental significance of the first stage: Defining goals. This is the most important element of the description of a business process. The lack of agreement in this regard, or the imposition of such goals by the management of the organization itself, point to a lack of understanding in regard to the principles of process management, which in turn might point to the lack of sufficient process maturity within the organization, preventing it from successfully implementing process management, which would result in the implementation having to be halted.

The fulfillment of each of the above tasks could trigger additional tasks—cultural, organizational, and technical—with the intent of preparing the organization to implement process management. Their performance (e.g. training sessions, changes to internal regulations, or defining the principles of managing contacts with external experts) could prove essential to the success of implementing dynamic BPM.

4.7.1 The (re)Design Stage

The aim of the (re)Design stage is to design or update process descriptions in a way which is tailored to their nature, the process maturity level of the organization, and the skills of the knowledge workers.

In this stage, process descriptions are designed or updated with a focus on the data processed in their course, in accordance with the principles adapted in the organization. As has been demonstrated in Chap. 2, process identification should be performed with the use of:

- the knowledge of the employees;
- exploratory research (discovering not just the course, but the entire context of performance);
- the standard models for the field in which the organization operates.

For static processes and selected static processes with dynamic exceptions, the process descriptions created in the (re)Design stage will usually have the form of diagrams or flowcharts depicting the process as a sequence of events. The natural course of events is to then use them to prepare and present the process description notation, e.g. in the form of BPMN or Gantt diagrams. In the case of dynamic processes and some static processes with dynamic exceptions, the process descriptions prepared in this stage may have the form of diagrams or flowcharts, but they will usually take the form of e.g. checklists or tables depicting the process as a collection of possible tasks to be accomplished during process execution, with the option of adding further unforeseen tasks in accordance with one's level of privileges. In the case of all processes, the form of presentation may change in the course of performance itself. The specific form of presentation should respond to the requirements of process performers with a view to facilitating performance, including the option to add or skip certain tasks. For instance, static processes may take the form of a process diagram in the planning stage and a Gantt diagram, a process diagram, or a 3D model in the *ex-post* analysis stage.

In the case of implementing dynamic BPM, organizations should identify and make a thorough evaluation of the possibilities of eliminating or mitigating the risks of infringing on essential interests of particular employees and groups of employees as the result of the planned process improvement. As in the case of implementations of traditional business process management, employees who feel their interests are threatened might resort to covertly sabotaging, or even directly resisting, the ongoing implementation in order to prevent, or at least postpone, the elimination of their privileges or benefits (e.g. position, additional authorizations, or parts of their salaries). In the case of implementing dynamic business process management, failing to eliminate the risk of infringing on essential interests could result in the limitation or even the complete elimination of the engagement of process performers in the creation or documentation of improvements to the performed processes. As the end result, it may limit the scope of the implementation to its traditional, static counterpart, without adding the value offered by the dynamic tailoring of processes to the specific context of performance, as well as their integration with knowledge management.

In the course of a project pertaining to the implementation of a new strategy of Human Resource Management (HRM) in one of the leading capital groups in Poland, it turned out that the implementation of advanced IT solutions and the HR Business Partner model may result in the necessity to reduce over half of the positions tasked with performing processes pertaining to HR and payroll documentation. This risk has been identified as early as in the stage of preparing a model of the improved processes, before the widespread consultations, implementation, and training sessions began. In effect of risk-reducing actions with the aim of finding meaningful work for the employees at risk of downsizing, the employees were approached with the proposal of undergoing widespread training with a view to changing their scope of responsibilities. In effect, thanks to widespread support, including the active support of employees to be transferred from the HR and payroll department, the implementation of new IT solutions was successful. The newly implemented processes encompassed the HR Business Partner model, which saw the creation of a team with expert knowledge on the organization and performing a much larger number of HR processes, encompassing such issues as: HR controlling, planning employee development initiatives, and supporting managers in the performance of "soft" HR processes pertaining to their subordinates. The transformation process has been planned to span 6 years. It was assumed that in that time, employment in the capital group would drop by 16%, and as many as 42% in HR, while the existing spectrum of tasks pertaining to human resource management (HRM) (mostly transactional) will be considerably expanded with issues pertaining to the active support of the business strategy by the field of HR.

4.7.2 The Communication and Cooperation Stage

Similar to the Transfer of knowledge stage in process descriptions pertaining to the implementation of traditional business process management, the aim of the Communication and cooperation stage is the dissemination and verification of knowledge contained in process descriptions. It is usually performed with the participation of a broad range of employees with ongoing access to the implementation team and the information repositories maintained in the course of implementation. Implementations of dynamic business process management require two-way verification, as well as the creation and discovery of knowledge used in the course of performance. The IT tools used in the process should not only enable access (in accordance with individual privileges) to repositories of information and data about current and past processes, but also allow for the day-to-day exchange of information within a community of practitioners (managers, experts, and employees) in accordance with the organizational culture, or at least group of employees drawing on information from similar fields of knowledge in their work.

4.7.3 The Implementation and Adjustment Stage

The aim of the Implementation and adjustment stage is to prepare the organization for the broadest possible use of its intellectual capital through the empowerment and constant motivation of its employees and the support of IT systems tailored to the requirements of the knowledge workers.

In this stage, the primary goal of the organization's management and the implementation team is to initiate changes to the organizational culture. Implementing dynamic business process management requires the rejection of the organizational silo culture, in which the actions of particular employees are the result of the orders of their supervisors, and empowering employees to act upon their knowledge in accordance with the goals of the process at hand. However, in the case of dynamic BPM, it is not enough to break down organizational silos and focus processes on the need of the clients instead of the organization's bureaucratic needs or quarterly reports—it is also essential to initiate the process of knowledge dissemination and define the practical methods and limitations (tailored to the employees' skills) of creating and improving processes (in the form of limited experiments).

Because—as has been noted in the *Introduction*—the implementation of dynamic BPM is impossible without the support of IT, in this stage of the implementation, prototypes of process-driven applications should be prepared, which enable the verification and potential reporting of improvements by future users in the scope of:

- the design of the user interface;
- the information content (data) of the user interface;

- the method (methods) of automated signaling to both stationary and mobile devices of information essential to process execution (e.g. required tasks, going over warming and alarm thresholds, received messages);
- available standard reports and analyses;
- integration with other used programs, including e-mail, messengers, and social media applications;
- integration with the organization's ICT infrastructure.

Subsequently, with the use of approved information, applications should be prepared with the aim of supporting process execution. Regardless of the type of processes in question, simulation research and prototyping should link process discovery and design with the automation of their performance. Work pertaining to process design and the preparation of applications supporting process execution may be performed by way of:

- workshops on the identification of tasks (needs) and introducing improvements to the processes performed in the organization;
- workshops on describing and modeling processes or ontologies describing the framework of executing case management—primarily with the participation of all employees engaged in the performance of the described processes;
- exploratory research (process discovery) using data collected in the course of process execution in transactional systems (ERP, CRM, dedicated domain-specific systems, etc.), as well as useful messaging software and social-media applications (social BPM);
- comparative research on the organization's processes or ontologies with process patterns specific to the field in question;
- simulation research for prepared process models (both static and dynamic) accounting for the actual data on the environment of process execution;
- prototyping and testing in BPMS systems with CMS functionality or CMS systems with BPMS functionality [44];
- validation and performance testing of finished process-driven applications and their integration with the organization's systems;
- accommodation of internal regulations with a view to maintaining the integrity of dynamic business process management with other fields of management.

4.7.4   The Execution and Monitoring Stage

The aim of the execution and monitoring stage is the efficient performance of work and the creative use of the knowledge and dynamism of the employees in order to create and offer innovative products, services, and process improvements. This stage also encompasses the monitoring of work performance and the management of knowledge related thereto, with the aim of learning by doing and the competitive use of verified or created knowledge.

Table 4.8 Comparison of the goals of process management and knowledge management in the performance and monitoring stage

Process-focused perspective	Knowledge-focused perspective
Efficient and successful performance of work	Creative use of the knowledge and dynamism of the employees
Management of work at a specific position	Management of knowledge verified and created at a specific position
Instant introduction of identified possibilities of improvements in order to create value or reduce risks	Rapid dissemination of knowledge in order to create value or reduce risks
Monitoring of work performance	Monitoring of knowledge management related to work performance
Focus on results as of today	Focus on maintaining relatively constant competitive advantage – the capability to use the intellectual capital of the organization with a view to creating value

Source Author's own elaboration

The performance and monitoring of dynamic BPM requires the strict, albeit flexible connection of work and knowledge, or, to be more precise, the connection of business processes and knowledge management (Table 4.8).

In accordance with the 2nd principle of dynamic BPM, this stage allows for holding limited experiments (attempts at improving processes). Changes to the organizational culture initialized in the Implementation and improvement stage should lead to the use of the entire dynamism of individual knowledge workers and their teams during process improvement. Without a culture in which workers have trust in and are not afraid to take the risk of holding limited experiments, it is impossible to successfully implement dynamic business process management.

The employees:

- due to idleness, fear of the consequences of deviating from the standard, or simply out of the reluctance to "overshadow" their supervisors will not introduce improvements, which breaks the 2nd principle of dynamic BPM,

  or

- out of the fear of being punished for deviating from standard processes, of being accused of exceeding their powers, will hide the fact of introducing improvements to the process (the so-called "hidden factory" effect), which breaks the 3rd principle of dynamic BPM,

  or

- out of the fear of the organization taking over their revealed, individually constructed knowledge and, in consequence, being let go from work, as well as due to internal competition, will hide the fact of introducing improvements to the process (the so-called "hidden factory" effect), which breaks the 3rd principle of dynamic BPM.

As has been noted in the introduction to this chapter, the implementation of dynamic BPM is impossible without the support of IT. In this stage, this support should ensure the ongoing monitoring and management of automated processes, including the most automated possible analysis of the entered improvements and their results on the basis of the full context of performance. Only then, given the large number of concurrent process executions, will it be possible to quickly and systemically identify innovations, collect new knowledge, reject outdated knowledge, and disseminate knowledge throughout the organization.

4.7.5   The Analysis and Diagnosis Stage

Process execution in accordance with the concept of dynamic BPM enables their performers to accommodate the course of the process to the requirements of the specific situation at hand. The automatic logging of such information in an event log of a BPMS or CMS system on a level 4 or minimum 3 of maturity allows us to reconstruct the course and the full context of performance [40]. In effect, by analyzing data from multiple process executions, it is possible to establish:

- contextual scenarios of process execution (e.g. a sales process dependent on the type of goods, the nature of the client, or the time of day; an investment process dependent on the nature of the investment, e.g. real estate, infrastructure) [45];
- experts in specific scopes of process execution (e.g. finance, production) and selected scopes of process management (e.g. optimization of efficiency or innovation) [46];
- process patterns for elementary subprocesses (e.g. decision subprocesses), which may be used in multiple processes [22];
- pattern indicators and groups of indicators exceeding simple statistics;
- new elements/criteria influencing process efficiency, requiring codification and needing to be taken into account in the further management of knowledge and processes in the organization.

As has been shown in the Sect. 3.4 of Chap. 3, this is an ongoing source of up-to-date practical knowledge, more essential than explicit knowledge or tacit knowledge collected from outside the organization. In accordance with the definition of knowledge management, it should be used to help achieve the organization's goals [47].

4.8 Conclusions

As has been discussed in this chapter, in the practical dimension not only does the concept of dynamic BPM enable the modeling of, but also simulation research on dynamic processes.

The use of process exploration in connection with simulation research on dynamic processes allows for considerable leaps in the efficiency and results of the processes in question, as well as the considerable reduction of risk during their implementation and the optimization of resources dedicated to their performance. Thanks to the discovery-driven analysis of deviations from the standard course of the process and the most common paths of process adaptation, it is possible to practically define the set of potential, implementable improvements and data which both directly and indirectly influence process execution. Their use in simulation research allows us to design new standard processes "as of today," as well as to define the resources needed in the course of their performance. Subsequent process executions and the analysis of subsequent dynamically introduced changes, and following that: subsequent process improvements and simulation research on their results enable the creation of a process improvement cycle, permanently combining the practical performances of dynamic processes with the simulation research focused on their optimization. Subsequent studies, including studies planned by the author, will investigate the extent to which this new cycle of improving dynamic processes will be in the KE quicker and more efficient than the traditional Deming cycle.

In order to meet these requirements, process management should be implemented (as early as in the Process discovery stage) with regard of the fact that process performers have at their disposal essential knowledge, which forms the touchstone of the intellectual capital of the organization, as well as the fact that in order for this capital to have a value-adding effect on the organization's bottom line, it must be used in the day-to-day process execution—not just in the course of process discover or periodic process reviews or audits. This implementation should not be limited to the intellectual capital in the hands of the organization's management, but the broadest possible group of employees—the process performers. The key is to accept and make use of the fact that "*no asset has greater potential for an organization than the collective knowledge possessed by all its employees.*"[48]. This realization considerably extends the goal of implementing (dynamic) business process management, supplementing it with the practical benefits of integrating process management with knowledge management as the basis of the long-term competitive advantage of the organization.

References

1. Senge P (1990) The fifth discipline. The art and practice of the learning organization. Currency Doubleday, New York
2. Bitkowska A (2013) Zarządzanie procesowe we współczesnych organizacjach. Difin, Warszawa
3. Knudson G (2013) What is BPM? Retrieved from http://www.bpmleader.com/2013/07/29/what-is-bpm/ [4.11.2013]
4. Highsmith J (2009) APM: agile project management creating innovative products, 2nd edn. Addison Wesley
5. Alliance S (2016) Learn about scrum. Retrieved from https://www.scrumalliance.org/why-scrum [18. 08. 2017]
6. Thompson JR, Koronacki J, Nieckuła J (2005) Techniki zarządzania jakością od Shewarta do metody "Six Sigma". Akademicka Oficyna Wydawnicza EXIT, Warszawa
7. Mahal A (2010) How work gets done: business process management, basics and beyond. Technics Publications, LLC, USA
8. Nowosielski S (2012) Zarządzanie procesami. Retrieved from http://procesy.ue.wroc.pl/uploads/pliki/procesy/wyklady/ZPRnowosielskiWYKLAD.pdf [8.08.2017]
9. Szelągowski M (2004) Szczegółowość identyfikacji procesów i działań w zarządzaniu dynamicznymi procesami biznesowymi. Zeszyty Naukowe. In: Studia i Prace, vol 49. KZiF SGH, Warszawie, pp 114–128
10. Nowosielski S (2009) Modelowanie procesów gospodarczych w literaturze i praktyce. In: Prace Naukowe UE we Wrocławiu, vol 52. Wydawnictwo Uniwersytetu Ekonomicznego, Wrocławiu
11. OMG (2013) Business process model and notation (BPMN) v2.0.2. Retrieved from http://www.omg.org/spec/BPMN/2.0.2/PDF/ [3.04.2016]
12. Liker JK (2005). Droga Toyoty. 14 zasad zarządzania wiodacej firmy produkcyjnej świata (The Toyota way. 14 management principles from the world's greatest manufacturer). MT Biznes Ltd, Warszawa
13. Trkman P (2009) The critical success factors of business process management. Int J Inf Manage 30(2):125–134
14. Dabaghkashani AZ, Hajiheydari BN, Haghighinasab CM (2012) A success model for business process management implementation. Int J Inf Electron Eng 5(2), Sept 2012
15. Rosemann M, vom Brocke J (2010) The six core elements of business process management. In: Handbook on business process management 1. Springer
16. Pucher M (2012) How to link BPM governance and social collaboration through an adaptive paradigm. In: Fischer L (ed) Social BPM: work, planning and collaboration under the impact of social technology. Future Strategies Inc, Lighthouse Point, pp 57–76
17. Di Ciccio C, Marrella A, Russo A (2012) Knowledge-intensive processes: an overview of contemporary approaches? In: 1st International workshop on knowledge—intensive business processes (KiBP 2012), Rome, Italy, 15 June. Retrieved from http://ceur-ws.org/Vol-861/KiBP2012_paper_2.pdf [2.04.2016]
18. OMG (2014) Case management model and notation (CMMN) v1.0. Retrieved from http://www.omg.org/spec/CMMN/1.0/PDF [1.08.2017]
19. Kania K (2013) Doskonalenie zarządzania procesami biznesowymi w organizacji z wykorzystaniem modeli dojrzałości i technologii informacyjno-komunikacyjnych. Wydawnictwo Uniwersytetu Ekonomicznego w Katowicach, Katowice
20. Rothschadl T (2012) Ad-hoc adaptation of subject-oriented business processes at runtime to support organizational learning. In: Stary C (ed) S-BPM ONE—scientific research: 4th international conference, S-BPM ONE 2012, Vienna, Austria, 4–5 April 2012. Springer, Berlin
21. Knudson G (2014) BPM-based case management frameworks. Retrieved from http://www.bizflow.com/bpm-software/blog/BPM-based-case-manage-ment-frameworks-CMF [04.05.2015]
22. Belaychuk A (2011) ACM: paradigm or feature? Retrieved from http://mainthing.ru/item/401/ [21.01.2011]

23. Dufresne T, Martin J (2003) Process modeling for e-business. INFS 770—methods for information systems engineering: knowledge management and e-business. Retrieved from http://odesso.com/sites/default/files/documents/processmodeling.doc [21.02.2017]
24. Mello A (2014) How to avoid the 10 biggest mistakes in process modeling. Retrieved from http://www.bptrends.com/how-to-avoid-the-10-biggest-mistakes-in-process-modeling/ [17.02.2017]
25. Waszkowski R, Kiedrowicz M (2015) Business rules automation standards in business process management systems. In: Kubiak BW, Maślankowski J (ed) Information management in practice. Wydział Zarządzania Uniwersytetu Gdańskiego, Gdańsk, pp 187–200
26. Betz S, Eichhorn D, Hickl S, Klink S, Koschmider A, Li Y, Trunko R (2009) 3D representation of business process models. Retrieved from http://subs.emis.de/LNI/Proceedings/Proceedings141/gi-proc-141-005.pdf [2.06.2017]
27. Effinger P (2013) A 3D-navigator for business process models. In: La Rosa M, Soffer P (eds) Business process management workshops. BPM 2012. Lecture notes in business information processing, vol 132. Springer, Berlin, Heidelberg
28. State of Queensland (2017) Possible cardiac chest pain clinical pathway. Retrieved from https://www.health.qld.gov.au/__data/assets/pdf_file/0021/439131/sw574-chest-pain-pathway.pdf [14.01.2018]
29. Gabryelczyk R, Jurczuk A, Pęczkowski M (2016) Determinanty wyboru notacji modelowania procesów biznesowych. Roczniki Kolegium Analiz Ekonomicznych SGH, no. 40, pp 357–370
30. Price RJ, Shanks G (2004) A semiotic information quality framework. Retrieved from http://citeseerx.ist.psu.edu/viewdoc/download?doi=10.1.1.83.9817&rep=rep1&type=pdf [17.02.2017]
31. Wahl T, Sindre G (2006) An analytical evaluation of BPMN using a semiotic quality framework. Retrieved from http://ceur-ws.org/Vol-363/paper14.pdf [17.02.2017]
32. Gabryelczyk R, Jurczuk A (2015) Comparative analysis of business processes notation understandability. Information management in practice. Uniwersytet Gdański, Wydział Zarządzania, Sopot
33. Sobczak A (2013) Architektura korporacyjna. Aspekty teoretyczne i wybrane zastosowania praktyczne. Ośrodek Studiów nad Państwem Cyfrowym
34. Kemsley S (2010) Runtime collaboration and dynamic modeling in BPM: allowing the business to shape its own processes on the fly. Cut IT J 23(2):35–39
35. Szelągowski M (2014) Konsekwencje dynamic BPM. E-mentor 4(56):61–68. Retrieved from http://www.e-mentor.edu.pl/artykul/index/numer/56/id/1126 [21.09.2016]
36. Gottanka R, Meyer N (2012) ModelAsYouGo: (re-)design of S-BPM process models during execution time. In: Stary C (ed), S-BPM ONE 2012, LNBIP 104. Springer-Verlag Berlin Heidelberg, pp 91–105
37. Ultimus (2004) Adaptive discovery. Accelerating the deployment and adaptation of automated business processes. Retrieved from http://www.ultimus.com/download-the-adaptive-discovery-whitepaper
38. ISIS Papyrus (2016) Adaptive processes. Retrieved from http://www.isis-papyrus.com/e15/pages/business-apps/adaptive-case-management/adaptive-process.html [4.04.2016]
39. Hammer M (1999) Reinżynieria i jej następstwa—jak organizacje skoncentrowane na procesach zmieniają naszą pracę i nasze życie (Beyond reengineering. How the process-centered organization is changing our work and our lives). Wydawnictwo Naukowe PWN SA, Warszawa
40. IEEE Task Force on Process Mining (2012) Process mining manifesto. Retrieved from: http://www.win.tue.nl/ieeetfpm/doku.php?id=shared:process_mining_manifesto [02.04.2016]
41. Gawande A (2009) The checklist manifesto how to get things right. Metropolitan Books, New York
42. Swenson K (2010) Mastering the unpredictable: how adaptive case management will revolutionize the way that knowledge workers get things done. Meghan-Kiffer Press, Tampa (USA)
43. Krogstie J, Sindre G, Jørgensen H (2006) Process models representing knowledge for action: a revised quality framework. Eur J Inf Syst 15(2006):91–102

44. IBM (2012) Scaling BPM adoption: from project to program with IBM business process manager, 2nd edn
45. Chan N, Yongsiriwit K, Gaaloul W, Mendling J (2014) Mining Event Logs to Assist the Development of Executable Process Variants. In: Jarke M et al (eds), CAiSE 2014, LNCS 8484, Springer International Publishing Switzerland, pp 548–563
46. HandySoft. (2012) Dynamic BPM—the value of embedding process into dynamic work activities: a comparison between BPM and E-mail. Retrieved from http://www.bizflow.com/system/files/downloads/HandySoft%20-%20Dynamic%20BPM%20White%20Paper_0.pdf [2.04.2016]
47. Murray P, Myers A (1997) The facts about knowledge. Special report, 11. 1997
48. Kaplan RS, Norton DP (2004) Measuring the strategic readiness of intangible assets. Harv Bus Rev

Conclusion

Even upon integration with knowledge management, process management does not encompass management in the organization as a whole. The aim of the organization is not to manage processes or knowledge to perfection. Instead, the aim of management in the organization—including process management and knowledge management—is to achieve results in accordance with the strategic goals of the organization (in business, these goals usually translate to profit) and to gain the most lasting competitive advantage over the competition [1]. Process management is intended to ensure the most efficient creation of value from the resources available to the organization in the knowledge economy—primarily from the organization's intellectual capital. But similar to financial capital, intellectual capital, when left unused, degrades in value. In order to thrive and multiply, it must be put to work. One of the goals of process management in the knowledge economy is to enable the operation and development of intellectual capital.

Dynamic business process management is an extension of our outlook on the organization as a collection of resources connected by dynamically managed processes [2]. In the case of organizations operating in the knowledge economy, of extreme importance is the capability to maintain dynamic equilibrium in the pursuit of goals, which are also updated on the basis of informal actions and information flows, which are often collectively referred to by the term: *social BPM*. Organizations must have access to resources and the capability of peripheral vision in order to quickly use knowledge from just-emerging processes with a view to creating value.[1] At present, traditional process management pursues standardization

---

[1]Peripheral vision—all that is visible to the eye outside the central area of focus; side vision. (Retrieved from http://www.dictionary.com/browse/peripheral-vision [10.07.2017]).

The use of information derived from peripheral vision to anticipate the need for correcting strategies and tactics in order to avoid risks or build new competitive advantages requires the implementation of not just a systemic mechanism of observing even the smallest changes in the organization's environment, but first and foremost, the capability to implement changes faster than the competition [3].

© Springer Nature Switzerland AG 2019
M. Szelągowski, *Dynamic Business Process Management in the Knowledge Economy*, Lecture Notes in Networks and Systems 71,
https://doi.org/10.1007/978-3-030-17141-4

and centralized management of all of the processes in the organization, which not only does not help the organization to quickly react to change, but often goes as far as to thwart change to begin with. This is almost as if we wanted to consciously control processes in the human body which are beyond our reach, such as controlling our pulse or controlling our hand after we touch a hot object. This will prove futile, because evolution equipped us with entirely different, much more effective mechanisms. Sometimes, the pursuit of standardization and centralization in organizations is an unnecessary burden, one which not only delays the organizations' reaction time to ongoing changes, but also their ability to perceive changes as such.

The conclusions presented in this work are the result of both practice and theoretical reflection. As has been noted in the *Introduction*, the present book pertains to a comprehensive, multi-year sequence of studies performed by the author, which have been confronted with practical experience on an ongoing basis, and which at times pointed to new fields and research goals. Within the performed work, the author managed to provide answers to the fundamental research questions:

1. **What goals should be set before process management in the knowledge economy?**
   The implementation of dynamic BPM requires organizations to account for in practice that in the knowledge economy, intellectual capital is more important than tangible goods, including financial capital. The goal of implementing process management in integration with knowledge management cannot be limited to the optimization or raising the efficiency of the performed processes. A good practice in this regard is to define the goal as the broadest possible use of knowledge with a view to creating measurable value, but also to renewing and developing the organization's intellectual capital in order to build long-term competitive advantage. The goal of implementing dynamic BPM (as an extension of the goal of traditional business process management) may, in effect, be defined as the "constant rise in efficiency and the use of knowledge" or "the optimization of operations and the management of the entire knowledge available to the organization"—or in a similar way, depending on the strategic priorities of the organization.

2. **Can traditional process management be used in the knowledge economy?**
   Traditional business process management is well-suited in the management of static processes, for which it is possible to define the specific, detailed algorithm of performance prior to performance itself. These are e.g. processes regulated by law or production processes strictly defined by requirements deriving from objective laws (e.g. physical or chemical laws), or stemming from a variable project, the variability of which is limited to the choice of one of a number of predefined scenarios. The number of such processes is becoming increasingly smaller in the knowledge economy. At present, it is estimated that they comprise a mere 20–30% of all processes in the organization. In effect, traditional business process management cannot be used in the knowledge economy, as it

leaves out an entire 70–80% of all processes in the organization outside the scope of management.

3. **Is it possible to extend traditional business process management in order to tailor it to the requirements of the knowledge economy?**
This entire book is devoted to demonstrating that this is indeed possible. There is no need to create separate concepts and methodologies from scratch. If they are created nonetheless, this fact has two non-substantive causes:

   - sales need (new concepts—new chance of sales, e.g. RPA, AI);
   - lack of theoretical reflection on and excessive attachment to current methods of operation, mainly in the environment of the proponents of process management. Many practitioners and researchers find it hard to come to terms with the fact that a concept which is so successful requires a substantive update. For this reason, they are ready to turn a blind eye to facts, negate the need for modifications, and "push out" proponents of change outside of the process management field, forcing them to create seemingly new concepts, which are seemingly disconnected from the evolution of process management to date.

4. **Is it possible to systemically manage knowledge hidden in the processes performed in the organization?**
Within the concept of dynamic BPM, it is possible to reveal tacit knowledge hidden in the processes performed in the organization in a manner which is completely consistent with the nature of the knowledge as such, that is, revealing by doing. IT solutions such as social media applications, the Internet of Things, process mining, robotic process automation (RPA), or machine learning (ML) allow for the online discovery and distribution of information encompassing the full context of the description of the performed processes, without the need of burdening the performers themselves with additional responsibilities. In effect, this solution is fully compliant with the 3rd principle of dynamic BPM—it allows for the discovery, verification, and transfer of the discovered knowledge to all of the authorized parties in the organization.

5. **Is it possible to integrate (dynamic) business process management and knowledge management, including the management of tacit knowledge?**
As has been demonstrated in Chap. 3, it is possible to integrate (dynamic) business process management and knowledge management, including the management of tacit knowledge. Such an integration removes most of the "eleven deadly sins of knowledge management," resulting in knowledge management becoming a part of the day-to-day management in the organization instead of an additional activity. The compliance between knowledge management models and business process management models shown in the work enables organizations to build a cohesive organizational and IT solution, which makes use of the knowledge derived from process performance with a view to creating value for clients and the owners of the organization. From the perspective of knowledge management, dynamic BPM is a constant source of

practical knowledge. Its use enables the ongoing management of the entire knowledge in the organization, ensuring:

- the constant, institutionalized readiness to change—thanks to the day-to-day search for new solutions by a broad range of employees;
- the capability of discovering, creating, and verifying knowledge on an ongoing basis by drawing on the experience of broad range of employees;
- the ability to use and transfer tacit knowledge, which thus far had been available only to its owner;
- the ongoing use of created knowledge at a faster pace than the competition— thanks to the systemic use of collective knowledge in the course of process performance.

6. **Is it possible to describe dynamically managed business processes?**
As has been demonstrated in Chap. 4, it is possible to describe dynamically managed business processes, even those which are unpredictable in nature. In such circumstance, the process model takes the form of a "collection of tasks" comprising the process, which may (or may not) be performed in practice. The course of the process is decided by the knowledge workers, which in the near future may be supported in this task by elements of artificial intelligence. Only upon the conclusion of the process may its documentation show the process as a sequence of specific tasks. Of course, in the case of traditional process management, where the process performer (and the artificial intelligence providing support) cannot deviate from the described process, the process as a sequence of events may immediately be described in the form of a process model. After all, it cannot be changed in the course of performance itself.
Proponents of adaptive/dynamic case management have been describing and performing dynamic processes, including unpredictable processes, with the support of IT systems for the last 10, or perhaps even 15 years. Chapter 4 has shown that it is possible to model dynamic processes with the help of the presently most popular 2.0 version of the BPMN notation, as well as to draw on the experiences of researchers and practitioners of case management with a view to introducing changes to the next version of the BPMN notation, thanks to which implementations and operations will be made even easier [4].

7. **Is it possible to perform simulation and optimization research on dynamically managed business processes, including unpredictable processes?**
Dynamic studies on dynamically managed business processes may be held in two ways, that is:

- in the form of simulation studies, which are usually held prior to process performance;
- in the form of exploratory studies in all three types of research fields defined in the Process Mining Manifesto [5, p. 4]. They may be held with the use of data derived from completed or ongoing processes.

Not only does the subchapter *Simulations of and optimization research on dynamic processes* from Chap. 4 demonstrate that research on dynamic processes is possible, it also presents the results of studies held by the author in cooperation with Piotr Biernacki on the practical methods of their performance.

8. **Should a dynamic business process have the same form of description throughout its entire lifecycle?**
   Studies held since 2015 by the author in collaboration with, among others, Cezary Lipiński, show conclusively that the form of description of a dynamic business process should change in the course of its lifecycle. Such is the expectation of all the groups of users who participated in the study. In practice, this result points to the necessity of changing the principles of building the user interface in systems supporting business process performance, both in the case of BPMS and CMS.

9. **Is it possible to reunify process management with case management?**
   Despite the lack of theoretical recommendations against combining the methodologies of process management and case management, researchers are divided on this matter. The so-called "process-centric" researchers see case management as a separate methodology with a limited scope of application, while the so-called "case-centric" researchers see process management as a bureaucratic concept which hinders the work of the knowledge workers. However, the dominating voice in the conversation is that of the practitioners. And practice increasingly more often shows that organizations wish to avoid situations in which they are forced to choose between process management and case management or to implement and maintain two process systems at once. For this reason, vendors of IT systems—such as IBM, Software AG, PegaSystems, or BPMOnline—already at present allow for the use of elements from the realms of both process management (e.g. in accordance with BPMN notation) and case management (usually in the form of a checklist) to describe a single process. Because the unification is both expected by the users and sought after by the vendors of IT systems, and there are no substantive counterarguments against it, the case for unification seems to be settled. An element which will speak for the quality of new (or updated) BPMS and CMS systems is not their approach to the integration of process management and case management. Rather, it would seem that the source of considerable qualitative differences between BPMS and CMS system is their integration with knowledge management and the functional incorporation of the concepts of process mining, robotic process automation, machine learning, and elements of artificial intelligence.

The above responses to the research questions posed in this work do not create a simple sequence of insights which would define a linear road-map for the future development of process management. The answer to question 4 does not only derive from the answer to question 3, but also from the answer to question 2, and indirectly—to question 8. in turn, the answer to question 8 is connected with the answers to questions 6 and 7, and indirectly—with question 4, etc. It would seem, therefore, that they present a picture of a certain ecosystem (as mathematicians and physicists would say: a phase space), in which value is created thanks to the operation of dynamic business process management in the knowledge economy [6].

The concept of dynamic business process management (*dynamic BPM*), which has been developed by the author since 2004, is not the first attempt at overcoming the limitations of traditional business process management with a view to tailoring it to the requirements of the increasingly more hypercompetitive environment of the organization in KE. However, it is the first cohesive theoretical attempt at formulating a generalization of process management in accordance with observed and predicted business requirements and emerging IT solutions, which are quickly adapted in practice. The aim is not to create a niche, fragmented solution (such as case management), but to propose a concept combining all of the available and emerging possibilities and containing all of the verifiable predictions. Full recognition of the consequences of using the concept dynamic business process management in practice will require further studies—both theoretical and practical—and the concept of *dynamic BPM* itself and the implementation methodology created on its basis will undoubtedly be updated several times in the future. However, even today it is apparent that the predictions resulting from the theoretical analysis of process management are confirming themselves and are finding practical applications. This is attested to by, among others:

- The results of studies supporting previous theoretical predictions, e.g. the falsification of the thesis the description of a process should remain unchanged throughout process performance (as well as the falsification of a thesis often used by proponents of process management in discussions with proponents of case management that processes should be described in the form of graphical diagrams![2]);
- The direction of the development of IT systems supporting management in the organization, e.g. Gartner's 2017 report, in which in accordance with theoretical predictions for dynamic business process management it was attested to in practice that case management and the possibility of making changes in the application by users are obligatory components of BPMS systems *Anno Domini* 2017.
- Compliance with other methodologies used and developed energetically in strict cooperation with process management, or which from the theoretical standpoint are already a part of (dynamic) business process management, such as methodologies and IT tools pertaining to process mining, machine learning, or artificial intelligence.

Another considerable confirmation of the validity of the concept of dynamic BPM will take the form of the broad practical verification of the prediction stated in the research question: "Is it possible to integrate (dynamic) business process

---

[2]"It's time to take insight, innovation, and strategy into action. That's about people and dynamic human interactions, not predefined workflow/BPM" [7].

management and knowledge management, including the management of tacit knowledge?". Section 2.4 of Chap. 2 presents the effects of development work, thanks to which new generations of IT systems supporting process management now contain, in accordance with theoretical considerations, the combined, harmonized functionality of knowledge management, elements of expert systems, or artificial intelligence. In effect, the integration of process management and knowledge management is possible and in practice already provides specific benefits, among which the most important is the considerable acceleration of the processes of verifying, creating, revealing, and sharing knowledge, and, first and foremost, the wide-scale use of knowledge [8]. This allows organizations to quickly provide innovative, personalized products and services with much lower risk than the competition, which does not have access to comparable solutions.

The phenomena described in the book in the course of e.g. answering the posed research questions confirm the existence of another qualitative step in the development of process management. According to the author, this step undoubtedly deserves the name 4th stage of the development of process management. In accordance with the research hypotheses confirmed by answering the research questions, the feature which makes this stage distinct from the previous, 3rd stage of the development of process management, is the use of (dynamic) business process management with a view to the broadest possible creation of value from intellectual capital as the source of the organization's competitive advantage. As has been discussed in Chap. 2, just like in the case of the previous stages of the development of process management, the 4th stage has first become apparent through practical solutions. For this reason, practice will be decisive in its confirmation and popularization in business.

As well as in the preparation of the business environment for the 5th wave of development of process management.

As the saying goes: *Panta rhei!*

Glossary

The exceptionally rapid development of process management and its practical solutions, including the methodologies and supporting IT systems deriving therefrom, has led to the ongoing emergence of new, as well as reinterpretation of existing terms and definitions. This may lead to disorientation among researchers and practitioners of management alike. The underlying cause rests in multiple authors using the same terms in a different meaning or using different terms to refer to the same phenomena. In order to avoid misunderstanding, the glossary defines the conceptual framework of the terms used in this book, which are tied to the broadly-understood field of business process management [9–12].

Dynamism
Dynamism—energy, ability, resilience in action, mobility, dynamics; an approach presuming inner activity, changeability, and the constant movement of nature [13].

Intelligence

(1) The ability to learn or understand or to deal with new or trying situations, also the skilled use of reason.
(2) The ability to apply knowledge to manipulate one's environment or to think abstractly as measured by objective criteria (such as tests) [14].

M. Szelągowski, *Dynamic Business Process Management in the Knowledge Economy*, Lecture Notes in Networks and Systems 71,
https://doi.org/10.1007/978-3-030-17141-4

1. Business process management

Business process (BP)

The term *business process* refers to an ordered sequence of actions with a prede-fined goal [15].[3]

Static processes (repeatable, predictable, routine), or, to be more precise, busi-ness processes which may be managed in a static manner, are those business processes for which it is possible to determine the precise, detailed algorithms of operation—prior to execution itself. This group of processes includes e.g. processes strictly regulated by law or production processes executed on the basis of patents, licenses, or dependent on objective laws (e.g. physical or chemical in nature).

Dynamic processes (*ad hoc*, adaptive, individual, unstructured, unpredictable, emergent, agile), or, to be more precise, business processes which require dynamic management, are those processes for which it is impossible to determine the precise, detailed algorithms of operation prior to performance. Their performance is dependent on the individual context (e.g. individual client requirements) and the knowledge (including the experience) of the employee or the team of employees performing the process. This group of processes also includes e.g. diagnostic-therapeutic processes, crisis management processes, and investment processes.

Business process management (BPM)

Business process management is a philosophy of conducting business,[4] as well as the methodologies and information and communication technologies (ICT) used in the scope of identifying, improving, and performing business processes with a view to raising their efficiency and profitability [16].

---

[3]To quote Davenport and Short: "We define business processes as a set of logically related tasks performed to achieve a defined business outcome."

[4]Philosophy of holding business—this work follows Garth Knudson [16] in thinking that this is a practical philosophy, understood as a collection of principles pertaining to the method of con-ducting business, including management, with particular focus on the management of human resources.

Traditional business process management (traditional BPM)

Traditional (static) business process management is a concept of management, as well as the methodologies and ICT tools used in the scope of identifying, improving, automating, and measuring the results of the performance of business processes with a view to raising their efficiency and profitability—under the assumption that process performers are not authorized to introduce changes in the course of execution itself [17]. In the case of traditional business process management, processes should be performed[5] in accordance with a predefined description or model. This means that traditional BPM encompasses static processes. Descriptions (models) of such processes change over periods which are much longer than the duration of process performance, which allows us to assume that processes remain unchanged in the course of performance itself. In effect, is possible to improve the processes on the basis of standard mechanisms of business process improvement (BPI). Such mechanisms usually derive from Edward Deming's concept and are based on different modifications of the PDSA cycle[6] [18–20]. Their use is based on the initiatives of the management or the personnel of the organization, evaluated and either accepted, rejected, or met with silence from the management. Performers of business processes managed in accordance with traditional BPM are not authorized to introduce changes to processes in the course of their performance. In many organizations, work deviating from accepted business processes (process models, procedures, rules and regulations) is seen as a breach of discipline and may be sanctioned, even if such departures are in the best interest of the organization.

Dynamic business process management (dynamic BPM)

Dynamic business process management (*dynamic BPM*) is an extension of traditional business process management, in which process performers are authorized to introduce dynamic adaptations to the requirements of performance in the course of process performance itself. The implementation of dynamic BPM should be

---

[5]For the purposes of this work the author has adopted a division into creative and reproductive executions, which correspond to the English terms:

• *execution*—the reproductive, passive execution of a chain of commands prepared and transmitted prior to execution (analogous to e.g. the preparation of a computer program in the form of a chain of commands in an EXE file (EXE from *EXEcution*). Even when the employee (*executor, doer*) sees mistakes in the program, he or she cannot change the program in the course of execution. This would require making changes to the source code itself, another compilation, and executing the new version of the program.)

• *performance*—the creative, active execution of work with a view to achieving a specific goal. Of course, such work should still make use of prepared standards, diagrams, or guidelines, but its performance is determined by the performer (analogous to performing a jazz improvisation in accordance with musical canons and thanks to previous preparation, albeit in a creative manner resulting from i.a. the context of performance, e.g. the reaction of the public.)

[6]The PDSA cycle (the Deming cycle, the Deming wheel, the PDCA cycle) created by Edward Deming is a diagram of a cycle of ongoing operation an improvement in the organization, consisting of four steps: Plan–Do–Study–Act. In literature, it is also often referred to as the PDCA cycle: Plan–Do–Check–Act.

performed in such a manner that the performance of a process be equal with its documentation, including the documentation of all changes and innovations [21, 22].

Knowledge-intensive business processes (kiBPs)

Knowledge-intensive business processes are those processes, in which the expected value is generated only by drawing on the knowledge of its performers, who perform interrelated tasks requiring making decisions on the basis of their knowledge [23, 24].

Process owner

The individual responsible for the design of the process or a group of processes, including the definition of their goals, performance indicators, course, and responsibilities for performance [25].

Process manager

The individual responsible for the ongoing support of the performers of a process or a group of processes, including responsibility for its goals an results; having the right to allocate resources with a view to performing a process or a group of processes or to petition the management for their allocation [25].

Process description

The term *process description* will be understood as all forms of descriptions defining the goal, the performance indicators, and the responsibilities for process performers. The most common forms of process descriptions are: verbal description, structured description (e.g. in the form of a procedure), block diagram, table, checklist, flow diagram, process diagram, Gantt diagram, 3D process map [26–30, pp. 27–29]. It is possible to make use of different forms of description within a single process description.

Business process management systems (BPMS)

The term business process management systems (also: business process management suites) refers to integrated IT systems supporting the lifecycle of process performance and improvements—from process identification and communication in the organization, simulation research, optimization and performance (automation), up to process monitoring, analysis, and constant improvement [31–33].

2. Case management

Case management (CM)

Case management (CM), which is also known as *adaptive case management* (ACM) and *dynamic case management* (DCM), shall be understood as IT tools along with their methodologies, which allow for work management on the basis of both structured and unstructured information used in business processes performed in a manner which is completely dependent on the decisions of the knowledge workers performing the processes [34–37].

Case management systems (CMS)

Case management is supported with the use of **case management systems** (CMS) —applications dedicated to the management of complex, highly unstructured business processes performed in a manner which is fully dependent on the decisions of the knowledge workers performing the processes with the use of and access to data. Such solutions support workflow and cooperation in the scope of management based on shared data resources, both structured and unstructured [38].

3. Knowledge management

Knowledge

"Knowledge is a fluid mix of framed experience, values, contextual information, and expert insight that provides a framework for evaluating and incorporating new experiences and information. It originates and is applied in the minds of knowers. In organization, it often becomes embedded not only in documents or repositories but also in organisational routines, processes, practices, and norms" [39, p. 4].

For the purposes of this work, a narrower definition of knowledge is used, which is closer to the definition of specialist knowledge:

Knowledge is a skill (individual or collective) resulting from experience, which allows for the evaluation, interpretation, and use of information in the specific context of its acquisition [39, 40].

Explicit knowledge

Explicit knowledge, sometimes referred to as formal, available, or expressible knowledge, is characterized by the formalized acquisition of information, the description of the context of its use, as well as guidelines on practical experience— usually derived from the available information ("I know 'what'") and the context of its acquisition. In organizations, explicit knowledge often takes the form of codified knowledge available in the form of procedures, rules and regulations, quality standards, process models, databases, methodologies of management, technologies, and descriptions of production processes.

Tacit knowledge

Tacit knowledge, sometimes referred to as confidential, silent, or quiet knowledge, is often overlooked in organizations. It is owned by particular individuals or teams of individuals. It is derived from experience ("I know 'how'"), being brought up in a particular cultural network, innate skills, etc. Within teams, this knowledge manifests itself in the organizational culture, views and attitudes, cultural norms, means of interpersonal communication, etc. This type of knowledge is usually revealed and developed in specific contexts of operation, outside of which access thereto is more challenging or even outright impossible. This knowledge also contains intuitions, assumptions, and subjective judgments, the sources of which are often not logically and formally documented, or even expressed by their owners —individuals or teams. However, this in itself does not prevent the owners of this knowledge from acting upon it, verifying it, or sharing it [41].

Knowledge worker

An employee whose main source of capital is knowledge. Knowledge workers have high degrees of expertise, education, or experience, and the primary purpose of their jobs involves the creation, distribution, or application of knowledge [42].

Knowledge management (KM)

Knowledge management (KM) is the whole of processes that enable the creation, dissemination and use of knowledge to achieve the organization's goals [43].

Process-oriented knowledge management (pKM)

Process-oriented knowledge management is the part of knowledge management which pertains to the distribution, acquisition, accumulation, and transfer of knowledge in the organization with a view to designing and performing business process, as well as knowledge created and uncovered in the course of performance. The aim of this concept is to provide employees with the knowledge connected with their tasks in the scope of operational business processes in the organization, as well as to transfer knowledge used in the course of process performance [44, 45].[7]

Intellectual capital (IC)

Intellectual capital comprises knowledge, intellectual abilities, and predispositions, which may be turned into value. In terms of value, intellectual capital is defined as the difference between the market value and the accounting value of the enterprise. It is comprised of three parts, which the organization can use with a view to creating value:

- human capital—the expert knowledge, skills, competences, experience, and motivation of the personnel;
- relational capital (client capital)—relations with invested parties, including primarily clients and vendors;
- structural capital (organizational capital)—which consists of: intellectual capital (patents, copyrights, trademarks) as well as infrastructural assets (organizational culture, strategies, management processes, IT systems) [46–48].

Knowledge economy, knowledge-based economy (KE)

The terms *knowledge economy* or *knowledge-based economy* refer to an economy built on the creation of knowledge (treated as production), its further sharing, or distribution, and the practical use of both knowledge and information [49].[8]

Learning organization

A learning organization is one which is proficient in performing tasks pertaining to the creation, acquisition, and distribution of knowledge, as well as the modification of its behaviors in response to new knowledge and new experiences [51].

---

[7]"pKM is defined as the management function responsible for the regular selection, implementation and evaluation of process-oriented KM strategies that aims at supporting and improving an organization's way of handling knowledge in order to improve organizational performance" [44].
[8]"knowledge-based economies"—economies which are directly based on the production, distribution and use of knowledge and information [50].

4. Information techniques supporting process management

Information and communication technology (ICT)
This term refers to all technologies enabling the collection, processing, distribution of, and access to information. It clearly underlines the role of the integration of: communication, computers, mass memories and audiovisual systems, as well as software (systems software, midleware, and applications dedicated to enterprises and individuals), which allow their users to collect, process, distribute, and access information.

Internet of things (IoT)
The Internet of Things is an advanced network of devices connected via the Internet, which allows for mutual links between the devices, anywhere in the world, which creates value in terms of being online in all aspects of life [52].

Big Data
Information resources characterized by such a degree of quantity, variability, and pace of creation or change that they require specific technologies and analytical methods with the aim of turning them into information of knowledge in a form which creates value [53].[9]

Process mining
Process mining—techniques, tools, and methods used in the preparation of a process model, its verification, and development on the basis of data on the actual course of processes from event logs commonly accessible in modern IT systems of the following classes: EMR, HIS, MRPII, ERP, CRM, CCM, etc. [5, 54].

Robotic process automation (RPA)
Robotic process automation (RPA) is a class of software which allows for the automation of routine, usually mass-repeated business processes or tasks which are usually supported by several IT systems [55].

Machine learning (ML)
A field of IT which provides computer systems with the possibility of gradually increasing their ability to efficiently search for solutions to posed problems ("to learn") on the basis of analyzed data, without explicit programming [56].

Artificial intelligence (AI)
In general, a field of knowledge encompassing fuzzy logic, evolutionary calculations, neural networks, artificial life, and robotics. Artificial intelligence (AI) has two basic meanings:

---

[9]"Extremely large data sets that may be analysed computationally to reveal patterns, trends, and associations, especially relating to human behaviour and interactions" [53].

- Firstly, it means a common research field of computer science and robotics, in which development of systems performing tasks which require intelligence when performed by humans is a research goal;
- Secondly, it means a feature of artificial systems which allows them to perform tasks that require intelligence, when made by humans [57].

References

1. D'Aveni R (1994) Hypercompetition managing the dynamics of strategic maneuvering. The Free Press, New York
2. Pucher M (2012) How to link BPM governance and social collaboration through an adaptive paradigm. In: Fischer L (ed) Social BPM: work, planning and collaboration under the impact of social technology. Future Strategies Inc, Lighthouse Point, pp 57–76
3. Szelągowski M, Lipiński C (2014) Organizm i organizacja. In: Jasieński MW (ed) Innowatyka—nowy horyzont. Centrum Innowatyki WSB-NLU, Nowy Sącz, pp 115–128
4. Szelągowski M (2018) BPMN update proposal for non-expert users. In: Burduk A, Chlebus E, Nowakowski T, Tubis A (eds) Intelligent systems in production engineering and maintenance. ISPEM 2018. Advances in intelligent systems and computing, vol 835. Springer, Cham. https://doi.org/10.1007/978-3-319-97490-3_64
5. IEEE Task Force on Process Mining (2012) Process mining manifesto. Pobrano z http://www.win.tue.nl/ieeetfpm/doku.php?id=shared:process_mining_manifesto
6. Heller M (2012) Filozofia przypadku. Copernicus Center Press, Kraków
7. Fingar P (2007) The greatest innovation since BPM. Retrieved from http://www.bptrends.com/publicationfiles/SIX-03-07-COL-TheGreatestInnovationSinceBPM-Fingar-Final.pdf [2. 12.2017]
8. Pucher M, Ruhsam C, Kim T, Kobler M, Mendling J (2014) Towards a pattern recognition approach for transferring knowledge in ACM. In: 2014 IEEE 18th international enterprise distributed object computing conference workshops and demonstrations
9. Elzinga D, Horak T, Chung-Yee L, Bruner C (1995) Business process management: survey and methodology. IEEE Trans Eng Manag 42(2)
10. Puah P, Tang N (2000) Business process management, a consolidation of BPR and TQM. In: Proceedings of the 2000 IEEE international conference on management of innovation and technology, 2000, vol 1. ICMIT 2000
11. Khan RN (2004) Business process management: a practical guide. Meghan-Kiffer Press, Tampa
12. Chang J (2006) Business process management systems: strategy and implementation. Auerbach Publications; Taylor & Francis Group, New York
13. Oxford Dictionary (2018) Dynamism. Retrieved from https://en.oxforddictionaries.com/definition/dynamism [11.06.2018]
14. Webster Dictionary (2018) Intelligence. Retrieved from https://www.merriam-webster.com/dictionary/intelligence [20.05.2018]
15. Davenport T, Short J (1990) The new industrial engineering: information technology and business process redesign. Sloan Manag Rev 31(4):11–27

© Springer Nature Switzerland AG 2019
M. Szelągowski, *Dynamic Business Process Management in the Knowledge Economy*, Lecture Notes in Networks and Systems 71,
https://doi.org/10.1007/978-3-030-17141-4

16. Knudson G (2013) What is BPM? Retrieved from http://www.bpmleader.com/2013/07/29/what-is-bpm/ [4.11.2013]
17. Gartner IT Glossary (2016) Business process management BPM. Retrieved from http://www.gartner.com/it-glossary/business-process-management-bpm [3.03.2016]
18. Deming WE (1986) Out of the crisis. Massachusetts Institute of Technology, Center for Advanced Engineering Study, Cambridge
19. Davenport T (1996) Some principles of knowledge management. Strategy Bus 1(2):34–40. Retrieved from https://www.strategy-business.com/article/8776?gko=f91a7 [10.11.2017]
20. Pande P, Neuman R, Cavanagh R (2003) Six Sigma. K.E. Liber S.C, Warszawa
21. Szelągowski M (2014) Konsekwencje dynamic BPM. E-mentor 4(56):61–68. Retrieved from http://www.e-mentor.edu.pl/artykul/index/numer/56/id/1126 [21.09.2016]
22. Gartner IT Glossary (2016) Dynamic business process management. Retrieved from http://www.gartner.com/it-glossary/dynamic-business-process-management-bpm [3.03.2016]
23. Gronau N, Müller C, Korf R (2005) KMDL—capturing, analysing and improving knowledge-intensive business processes. J Univers Comput Sci 4(11)
24. Di Ciccio C, Marrella A, Russo A (2015) Knowledge-intensive processes characteristics, requirements and analysis of contemporary approaches. J Data Semant 4(1):29–57. Retrieved from https://www.researchgate.net/profile/Claudio_Di_Ciccio/publication/269629902_Knowledge-Intensive_Processes_Characteristics_Requirements_and_Analysis_of_Contemporary_Approaches/links/576a501a08ae1a43d23bca3c.pdf [18.07.2017]
25. Jeston J, Nelis J (2013) Business process management. Practical Guidelines to successful implementations, 3rd edn. Butterworth-Heinemann, Oxford
26. Dufresne T, Martin J (2003) Process modeling for e-business. INFS 770—methods for information systems engineering: knowledge management and e-business. Retrieved from http://odesso.com/sites/default/files/documents/processmodeling.doc [21.02.2017]
27. Betz S, Eichhorn D, Hickl S, Klink S, Koschmider A, Li Y, Trunko R (2009) 3D representation of business process models. Retrieved from http://subs.emis.de/LNI/Proceedings/Proceedings141/gi-proc-141-005.pdf [2.06.2017]
28. Effinger P (2013) A 3D-navigator for business process models. In: La Rosa M, Soffer P (eds) Business process management workshops. BPM 2012. Lecture notes in business information processing, vol 132. Springer, Berlin, Heidelberg
29. Mello A (2014) How to avoid the 10 biggest mistakes in process modeling. Retrieved from http://www.bptrends.com/how-to-avoid-the-10-biggest-mistakes-in-process-modeling/ [17.02.2017]
30. Gawin B, Marcinkowski B (2013) Symulacja procesów biznesowych. Helion, Warszawa
31. OMG (2011) Business process model and notation (BPMN). Retrieved from http://www.omg.org/spec/BPMN/2.0 [3.04.2016]
32. Kania K (2013) Doskonalenie zarządzania procesami biznesowymi w organizacji z wykorzystaniem modeli dojrzałości i technologii informacyjno-komunikacyjnych. Wydawnictwo Uniwersytetu Ekonomicznego w Katowicach, Katowice
33. Gartner IT Glossary (2016c) Business process management suite BPMS. Retrieved from http://www.gartner.com/it-glossary/bpms-business-process-management-suite [3.03.2016]
34. Swenson K (2010) Mastering the unpredictable: how adaptive case management will revolutionize the way that knowledge workers get things done. Meghan-Kiffer Press, Tampa (USA)
35. Swenson K (2010) Comparison: ACM vs. BPM. Retrieved from http://www.xpdl.org/nugen/p/adaptive-case-management/public.htm [20.02.2016]
36. Belaychuk A (2011) ACM: paradigm or feature? Retrieved from http://mainthing.ru/item/401/ [21.01.2011]
37. Aalst W, Weske M, Grunbauer D (2005) Case handling: a new paradigm for business process support. Data Knowl Eng 53(2005):129–162
38. Gartner IT Glossary (2016d) Case management solutions. Retrieved from http://www.gartner.com/it-glossary/case-management-solutions [3.03.2016]

39. Davenport T, Prusak L (1998) Working knowledge—how organisations manage. What they know. Harvard Business School Press, Boston
40. Tiwana A (2001) The essential guide to knowledge management: e-business and CRM applications. Prentice-Hall, Upper Saddle River
41. Fazlagić J (2006) Zarządzanie wiedzą. Szansa na sukces w biznesie. Gnieźnieńska Wyższa Szkoła Humanistyczno-Managerska, Gniezno
42. Davenport T (2005) Thinking for a living. How to get better performance and results from knowledge workers. Harvard Business School Press, Boston
43. Murray P, Myers A (1997) The facts about knowledge. Special Report, 11.1997
44. Maier R, Remus U (2003) Implementing process-oriented knowledge management strategies. J Knowl Manag 4(7):62–74
45. Woitsch R, Karagiannis D (2005) Process oriented knowledge management: a service based approach. J Univers Comput Sci 11(4):565–588
46. Edvinsson L, Malone M (1997) Intellectual Capital. Harper Collins Publishers, New York
47. Beyer K (2012) Kapitał intelektualny jako podstawa przewagi konkurencyjnej przedsiębiorstw. Uniwersytet Szczeciński. Studia i Prace WNEiZ US no. 25
48. Kianto A, Ritala P, Spender J, Vanhala M (2014) The interaction of intellectual capital assets and knowledge management practices in organizational value creation. J Intellect Capital 3 (15):362–375
49. Trewin D (2002) Measuring knowledge-based economy and society. An Australian framework. Canberra, Australian Bureau of Statistics, 2
50. OECD (1996) The Knowledge-based economy. General Distribution OCDE/GD(96)102
51. Garvin D (1998) The processes of organization and management. Sloan Management Review, 15 July 1998. Retrieved from http://sloanreview.mit.edu/article/the-processes-of-organization-and-management/ [15.04.2016]
52. Want R, Schilit B, Jenson S (2015) Enabling the internet of things. Computer 48(1):28–35
53. Oxford Dictionary (2018) Big data. Retrieved from https://en.oxforddictionaries.com/definition/big_data [11.06.2018]
54. Aalst W (2016) Process mining—data science in action, 2nd edn. Springer, ISBN 978-3-662-49850-7, pp 3–452
55. Software AG (2018) Four reasons why leading companies are implementing RPA
56. Oxford Dictionary (2018) Machine learning. Retrieved from https://en.oxforddictionaries.com/definition/machine_learning [11.06.2018]
57. Flasiński M (2016) Introduction to artificial intelligence. Springer, Switzerland